STATISTICS FOR
TOXICOLOGISTS

DRUG AND CHEMICAL TOXICOLOGY

Series Editors

Frederick J. DiCarlo
Science Director
Xenobiotics Inc.
Denville, New Jersey

Frederick W. Oehme
Professor of Toxicology, Medicine and Physiology
Director, Comparative Toxicology Laboratories
Kansas State University
Manhattan, Kansas

Other Volumes in Preparation

STATISTICS FOR TOXICOLOGISTS

David S. Salsburg
Pfizer Central Research
Groton, Connecticut

CRC Press
Taylor & Francis Group
Boca Raton London New York

CRC Press is an imprint of the
Taylor & Francis Group, an **informa** business

First published 1986 by Marcel Dekker, Inc.

Published 2019 by CRC Press
Taylor & Francis Group
6000 Broken Sound Parkway NW, Suite 300
Boca Raton, FL 33487-2742

© 1986 by Taylor & Francis Group, LLC
CRC Press is an imprint of Taylor & Francis Group, an Informa business

First issued in paperback 2019

No claim to original U.S. Government works

ISBN 13: 978-0-367-45160-8 (pbk)
ISBN 13: 978-0-8247-7590-2 (hbk)

Visit the Taylor & Francis Web site at
http://www.taylorandfrancis.com

and the CRC Press Web site at
http://www.crcpress.com

Library of Congress Cataloging in Publication Data

Salsburg, David, [date]
　Statistics for toxicologists.

　(Drug and chemical toxicology ; v. 4)
　Bibliography: p.
　Includes index.
　1. Toxicology—Statistical methods. 2. Toxicology, Experimental—Statistical methods. I. Title. II. Series: Drug and chemical toxicology (New York, N.Y. : 1984) ; v. 4.
RA1199.4.S73S25 1986　　615.9'0072　　86-4387
ISBN 0-8247-7590-2

Dedicated to my parents,

Frances and Theodore Salsburg,

who taught me the values

and morality of human life

by word and example

About the Series

Toxicology has come a long way since the ancient use of botanical fluids to eliminate personal and political enemies. While such means are still employed (often with more potent and subtle materials), toxicology has left the boiling-pots-and-vapors atmosphere of the "old days" and evolved into a discipline that is at the forefront of science. In this process, present-day toxicologists adopted a variety of techniques from other scientific areas and developed new skills unique to the questions asked and the studies being pursued. More often than not, the questions asked have never been thought about before, and only through the advances made in other disciplines (for example, in analytical chemistry) were the needs for answers raised. The compounding concerns of society for public safety, the maintenance of environmental health, and the improvement of the welfare of research animals have expanded the boundaries in which toxicologists work. At the same time, society has spotlighted toxicology as the science that will offer the hope of safety guarantees, or at least minimal and acceptable risks, in our everyday chemical encounters.

This *Drug and Chemical Toxicology* series was established to provide a means by which leading scientists may document and communicate important information in the rapidly growing arena of toxicology. Providing relevant and forward-looking subjects with diverse and flexible themes in an expedited and prompt publication format will be our goal. We will strive in this vehicle to provide fellow toxicologists and other knowledgeable and interested parties with appropriate new information that can be promptly applied to help answer current questions.

The application of statistical methods to assist in answering these relevant questions is a powerful tool—and yet it is a confusing and often little understood discipline for the toxicologist who is accustomed to working with whole animals, biological systems and fluids, or chemical and biochemical entities. To have a nontoxicologist push numbers into a formula that apparently was selected arbitrarily, generate figures (often to two or three decimal places!), and then pronounce that the results of two and one-half years' work was "not significant" may have a ring of heresy to the ears of listening toxicologists!! It is to bring these two positions into harmony that David Salsburg wrote *Statistics for Toxicologists* and that we are excited to include this volume as the fourth in the *Drug and Chemical Toxicology* series. Dr. Salsburg has divided his discussion into three areas: the acute toxicity tests (LD_{50}); subchronic toxicity studies; and tests of chronic toxicity. In each area he examines the statistical basis of the toxicity studies, reviews the lessons that he has learned from evaluation of the procedure, and then delves into the theories employed in applying statistical procedures to each of these tests. The result is a working statistician's appraisal of what conclusions each of the toxicity tests allow and how toxicologists can best apply statistical methods to derive more confidence from experimental observations. What more insights can the researcher ask for?! We hope you enjoy *Statistics for Toxicologists* as much as we did and, of course, as David Salsburg did in preparing it.

Frederick W. Oehme
Frederick J. DiCarlo

Preface

This book is dedicated to my parents, who are in their 80s and still living together. They taught me that to preserve the dignity of individual human beings is essential to civilization. That dignity means being able to grow old gracefully, free from the parasites that prey on wild species, with access to good food and water, protected from mycotoxins and not forced to satisfy hunger on spoiled foods. That dignity means freedom from the racking pains and fever of disease, protection from deadly epidemics, and easing of the chronic pains of old age. They taught me that each person has a right to respect and dignity. I have been working in biostatistics because I believe that we can use technology to make that possible.

The fact that so many of my parents' generation have reached their age was not predictable from past patterns of human demography. Although the modern world has seen an increasing life span over the past 100–150 years, this has been due mainly to a decrease in infant mortality and deaths in early adulthood. Whether we examine ages at death from Roman tombstones, headstones in medieval English cemeteries, or from the early years of this century, the patterns of life expectancy past age 60 have always been the same.

This is no longer true. My parents' generation is the first to benefit from the applications of sophisticated chemistry to life. They have had anti-infective agents to protect them from the diseases of early and middle life and drugs to ameliorate the problems of the chronic diseases in old age. Their generation was the first to eat a wide variety of wholesome foods, fortified with essential vitamins and minerals and protected from the toxins produced by microflora and microfauna.

This could not have happened without the science of toxicology. We can never expect to predict all the subtle effects of chemicals on the human environment by using animal models, but we have succeeded in identifying a large number of chemicals that can be used as drugs and preservatives and that, in agricultural use, are less dangerous than others at the expected levels of exposure. For all the uncertainties that remain, modern toxicology has succeeded in clearing chemicals without which my parents' generation would have had patterns of death past age 60 no different from those recorded on Roman tombstones.

This book is presented in the humble recognition that, with or without statistics, toxicology has worked.

David S. Salsburg

Introduction

This book assumes a mathematical background no higher than high school algebra and no prior knowledge of statistics; however, it does assume that the reader has met some of the statistical models in use in toxicology. I have also avoided detailed derivations of statistical algorithms and usually describe them in words rather than in formal algebraic structures, because I expect that the reader will have access to computer programs with which to run calculations. The major thrust of this book is to explain how these programs work and why and when they can be applied to problems in toxicology.

The view of toxicology discussed in this book is limited to the standard protocols used to clear products with regulatory bodies. The majority of articles appearing in the toxicological journals deal with mechanisms of toxic activity for specific compounds, often derived through elaborate and impressive mathematical-chemical models. However, since it has been my experience that the majority of toxicologists spend most of their time running and interpreting studies from these standard protocols, it is the purpose of

this book to show how statistical models are involved in the interpretation and design of such studies.

    This book contains three units. Each unit deals with a specific
class of protocols and the attendant statistical methods. The first
chapter of each unit discusses the statistical models used and
their applications in a general fashion. The second chapter discusses problems that can arise from the blind use of the statistical
models and establishes warnings on what the underlying assumptions of the statistical model may mean in terms of the scientific
interpretation of the results. The third and last chapter of each
unit is a detailed discussion of the statistical theory associated
with the methods and models that were presented in the two previous chapters. Chapters 1, 4, and 7 provide a unified over-
view of statistical methods as they can be and are applied to these
protocols. Chapters 2, 5, and 8 give an understanding of how to
apply enlightened skepticism to these methods. Chapters 3, 6, and
9 cover all the material usually discussed in two semesters of elementary statistics.

Contents

Contents

STATISTICS FOR
TOXICOLOGISTS

UNIT I
The LD$_{50}$

1

The LD$_{50}$

First Principles

What is the dose that kills? This may be a fundamental question
in toxicology, but it has no simple answer. In acute toxic studies we notice that the dose that may kill one animal will not kill
another of the same species and strain, and so we conceive of
each animal as having an associated lethal dose. If we draw at
random from the pool of all animals of a particular strain and
species available to us, we might get an animal that will be
killed by a single dose of 5 mg/kg or we might get an animal
that will be killed by a single dose of 5 g/kg. However, we
observe that most animals of a given strain will be killed by a
narrow range of doses, in spite of the very wide range that is
theoretically possible.

If we were to plot the proportion of animals that would be
killed by a given dose x or less on a graph as in Fig. 1, we
would see that very few animals will be killed by a low dose, so
the curve starts flat and near zero. At a much higher dose

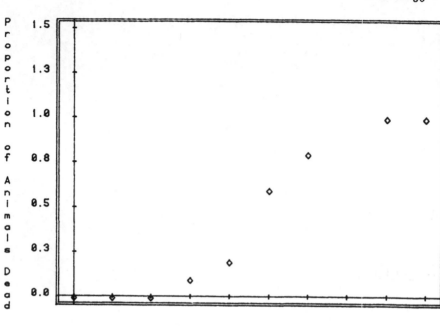

Figure 1   Typical LD$_{50}$ experiment.

there is a precipitous rise in the number of animals that would be killed, until we reach a dose that will have killed off most of them.   At that point the curve flattens out again since the maximum probability of killing is 1.0.   This *S*-shaped curve is called a *sigmoid*, which merely means that it is shaped like an *S*.   We can describe this graph in mathematical terms in the following fashion:

Probability of death = some algebraic expression of dose

or

P(death) = f(dose)

The expression *f(dose)* is a convenient shorthand.   It is pronounced "*f of dose.*"   It states that there exists some sort of a mathematical expression that will be eventually used, either on a computer or later in the discussion, but we are ignoring the specific form of that expression for the moment to concentrate on the more general idea. This is called *functional nota-tion* and it is widely used in mathematics. Its purpose is to avoid having to think about the details of the mathematics and algebra while thinking about something else. We take for grant-ed that somewhere, somehow there is a detailed specification of this function. At any point we put into functional notation only those concepts we need for the immediate discussion.

For instance, we may want to discuss the possibility that the probability of death for a given dose is different for males than it is for females. Then we would write

P(death) = f(dose,sex)

This is sufficient to show that there are different expressions for sex, just as there are different expressions for different doses. The components dose and sex are called the *arguments* of the function. There may be other arguments that we are not indicating at a particular moment because they are not germane to the current discussion.

Concentrating on the probability of death as a function of only one argument, dose,

P(death) = f(dose)

as shown earlier, we can indicate something about the mathe-matical form of the function. Because it is a probability, that is, a number that lies between zero (impossible event) and unity (certain event), we know that the plot of the curve will be a sigmoid. What else can we know?

Let us suppose that it were possible to determine the exact dose that will kill for any one animal by some laboratory tech-nique. Suppose further that we are able to collect a very large sample of animals. Then we could simply plot the percentage of animals that we observe being killed at each dose and obtain a curve as in Fig. 2. That set of points would be adequate to describe the function

f(dose)

This may not have the intellectual elegance of a complicated al-gebraic formula, but it would be adequate to describe the func-tion, and we would know the "dose that kills."

But we cannot do this for two reasons: First, there is no way to identify the exact dose that will kill each animal and,

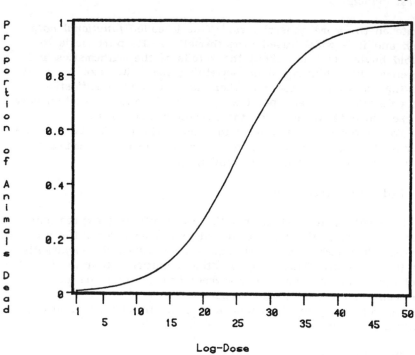

Figure 2   Theoretical LD curve.

second, it is not possible to collect a large enough group of animals in an experiment to be sure we have an accurate picture of the probabilities involved. Because of these two limitations, we are forced to collect data that bear only indirectly on our problem and we must therefore estimate the nature of the function from what data we can collect. The next sections deal with these problems.

Introducing Uncertainty

Since we cannot determine the exact dose that will kill a given animal, we pick a set of trial doses to use in an experiment. If an animal dies at a dose of, say, 35 mg, then we can assume that the exact dose needed to kill that animal would have been 35 mg or less. Say we administer a set of doses to 1000 mice and obtain the following table:

Dose (mg)	Number of mice dead	Percentage of mice dead (%)
5	0	0.0
15	11	1.1
30	50	5.0
60	120	12.0
120	553	55.3
240	890	89.0
480	995	99.5

We can be reasonably sure that 12% of the mice would have died at a dose of 60 mg or less, but we cannot determine what percentage would have died at a dose of 40 mg or what dose is needed to produce a death rate of 10%.

So the first source of uncertainty lies in the nature of what is experimentally possible. We have to estimate the points on the curve of f(dose) that we did not include in the experiment. We do this by requiring a mathematical formula for f(dose). We use the observed data to fit a formula taken from a small set of possible formulas. At this point the mathematics needed goes beyond the ordinary algebra that is all this book assumes of the reader. There is no simple algebraic formula that will produce a signmoid curve of the type we need for f(dose). One of the least complicated looks like this

$$P(\text{death}) = \sin\{[A + B(\text{dose})]^2\}$$

where A and B are constants. Other formulas involve indefinite integrals, logarithmic functions, and other paraphernalia found in calculus. However, just as we put aside the problem of what f(dose) looks like in algebraic notation in order to look at its overall structure, so too can we put aside much of the notational burden of calculus by considering a physical model of the curve.

Suppose we had drawn Figure 2 on a rubber sheet. Suppose further that we stretched the upper and lower edges of that rubber sheet until the curve in figure 2 became a straight line. The curve is already nearly straight in the center, so most of the stretching would have to be done in the regions where it flattens out. This "stretching" of the vertical axis is called a *transformation*. We transform the expression

P(death)

into something we symbolize as

W(P(death))

which we read "W of P of death." Then the function f(dose) can be written as a linear function, so that

W(P(death)) = A + B(dose)

[At this point there is an ambiguity in the notation. We have used the parentheses to denote an argument of a function, so f(dose) is read "f of dose." Now we are using the parentheses to indicate an argument on the left side of the equation, reading it "W of P of death," but we are also using the parentheses to indicate multiplication on the right-hand side. So the right-hand side means that we take a number symbolized by the letter A and add it to a number symbolized by the letter B which has been multiplied by the dose. Mathematics is not as rigorously unambiguous as it claims to be.]

Why is this called a linear function? Because, if you plot the formula

y = 5 + 3x

on a graph with y on the vertical axis and x on the horizontal axis, you get a straight line, as in Fig. 3. In general, the formula describes a straight line on the graph. When we transform the probability of death to get the straight line

W(P(death) = A + B(dose)

the number symbolized by B determines how rapidly the central portion of the sigmoid rises, and the numbers symbolized by A and B are used to determine the point associated with a 50% probability of death.

More Uncertainty

So far we have treated the sigmoidal curve of probability of death as if we could plot its points exactly from the data of an experiment. This requires that a very large number of animals

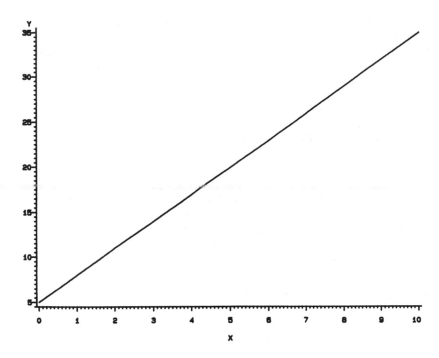

Figure 3 Graph of y = 5 + 3x.

be tested at each dose used, so that the proportion of animals observed to die at each dose would be an accurate estimate of the true probability of death associated with that dose. However, we cannot test enough animals in a real-life experiment to guarantee such accuracy. Instead, the usual LD_{50} experiment involves 5–10 animals at each dose.

Suppose we test 10 animals at a dose that has a true underlying probability of death of 50%. We cannot be assured that exactly 5 of the animals will die; owing to the random choice of animals, all 10 might have individual deadly doses greater than the test dose chosen. Or 9 of them might be resistant to that dose. Thus an actual experiment with 10 animals treated with a specific dose need not reflect the true underlying probability of death at that dose. But we need not assume that nothing can be concluded from such an experiment.

If the probability of death for a given dose is some small number, like 5%, it would seem highly improbable that all 10 animals would die. Thus, if we observe that all 10 animals die, then we can be reasonably sure that the underlying probability

of death for that dose is above 5%. How far above? The mathematical theories of statistics allow us to calculate a range of
values that are reasonable in the face of a given amount of evidence. Thus the fact that 10 out of 10 animals died can be
used to calculate that the probability of death is most likely
greater than 75%.

In addition to being able to determine reasonable bounds on
the underlying probabilaities of death at each dose tested, we
can use the mathematical formulation of

   f(dose)

to obtain even tighter bounds on the underlying probabilities.
In fact, we can use the same general techniques to create
three sigmoidal curves, one of them providing a "best-fitting"
curve, one an upper bound, and one a lower bound. Figure 4
shows three such curves for the following data:

Dose (mg)	Number of animals dead out of 10
5	0
10	1
20	1
40	5
80	4
160	8
320	10

Note that there is an apparent "dose reversal" from 40 to 80 mg.
This is due, of course, to the random noise of the experiment.
The fact that we know the true dose response is increasing
from one dose to the next is used in the calculation of the
curves in Figure 4.

The curves in Figure 4 have appeared as Minerva full grown
from Zeus's head. They can be calculated and plotted by any
of a number of standard computer programs. It is not the intent of this textbook to teach the student how to do those calculations. In order for the computer to make them, it was necessary to make certain assumptions about the nature of the
dose response, and it will be important for the toxicologist who
plans to use such programs to understand what these assumptions are and how they can or cannot be justified by the bio-

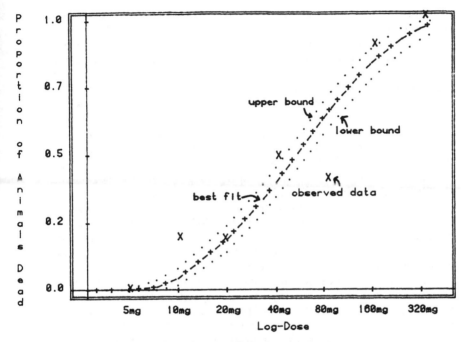

Figure 4 Sigmoidal fit with confidence bounds.

logy. In what follows we will examine these assumptions and
consider alternative mathematical models to those most widely
used in toxicology. The computer programs should be considered as "black boxes," much as one considers a television set:
The student needs only to know how to run the black box, how
to interpret its output, and how to apply enlightened skepticism
to that output, and not how to examine the algebraic manipulations involved in the program.

Models of Sigmoid Curves

The most widely used mathematical model for the sigmoid curve of

$$P(\text{death}) = f(\text{dose})$$

is the probit curve. The original idea for the probit curve
(along with its name) is due to C. I. Bliss, an American statistician who was an entomologist at the U. S. Agricultural De-

partment during the 1930s. Bliss was working with insecticides, trying to determine the doses that would be best for field use. He would expose a large number of insects to a given dose of poison in the lab and count the number that died. He noticed that no matter how low a dose he used, there were always some insects dead after exposure, and no matter how high a dose he used, there were always some insects still alive after exposure. He also noticed that when he had a dose that killed around 20% of the insects, it took a doubling of that dose to increase the probability of death by a fixed amount. That is, if he could kill 20% of the insects with a dose of 200 units, it would take 400 units to kill 40% and 800 units to kill 60%.

The mathematical model Bliss chose reflected these observations. He looked for a probability function that would never give a zero probability no matter how low the dose, or a 100% probability, no matter how high the dose. In view of the second phenomenon, Bliss sought a curve where the natural metameter of the dose would be the logarithm.

This last sentence should come as a shock to most students: The words *metameter* and *logarithm* are not in the vocabulary of the mathematical background presupposed for this text, so the student deserves an explanation. Those who know what these terms mean can skip the next three paragraphs.

When an animal ingests a toxin, the amount of material that reaches the ultimate site of action depends on more than just the dose given. Absorption, excretion, and metabolism of the toxin are influencing factors and therefore the amount of material delivered to the affected organ will be a function of the administered dose. We have no way of knowing exactly what that function is for most toxins, but we can approximate it by a smooth mathematical expression that seems to "work" most of the time. This smooth mathematical expression which is used to replace the measured dose is often called a metameter, a change in the units of dosing.

Bliss's observation about the effects of a doubling of dose is commonly made in biological research: Ever-increasing amounts of drug or toxin are usually needed for the same additive effect. A convenient metameter that reflects this fact is the logarithm. There are a number of ways to express the logarithm, but the most widely used is the "common logarithm." The following table displays the general pattern of the common logarithm:

Original dose (mg)	Logarithm of the dose
0.01	−2
0.10	−1
1.0	0
10.0	1
100.0	2
1000.0	3

Note that a dose between 1 and 10 mg (such as 8 mg) will have a logarithmic value between 0 and 1. Note also that the negative values of the logarithm can go to minus infinity: The logarithm of 0.001 is −3, that of 0.0001 is −4, and so on. There is no way of representing the logarithm of zero; the closer we get to zero, the more negative the logarithm.

This last property, the inability of the logarithm to represent zero, was convenient in Bliss's original formulation of a curve to model the effects of an insecticide. Bliss set up a model in which the dose killing a given insect was distributed according to the Gaussian law of probability where the units were the logarithm of the dose. A description of the Gaussian law and some justification for Bliss's choice are given in Chapter 3 of this book. For the moment, think of the Gaussian law as a "black box" method of stretching the ends of the sheet on which we sketch the curve of

P(death) = f(dose)

However, to understand how and when to use this particular black box, the student should recognize that Bliss's model (which he called *probit* to illustrate its use of "probability") has three important properties:

1. There is no dose for which one can get either a 0 or a 100% probability.
2. Though they never reach the points, the flat ends of the sigmoid curve go to 0 and 100% much more rapidly than for some other curves.
3. The sigmoid based on a Gaussian probability can be characterized entirely by two numbers: the dose associated with a 50% probability of death and the slope of the linear portion of the sigmoid.

The first two properties limited the way in which Bliss's probit model can be applied to biological data. The third property allows for mathematical convenience in calculating the points on the best-fitting curve or the upper and lower bounds on that curve.

Mathematical convenience was important in the 1930s, when Bliss proposed the probit, because all calculations had to be done with desk calculators and printed tables of advanced calculus functions. Probit tables were available at that time because the Gaussian distribution was widely used. Bliss (with the aid of R. A. Fisher) was thus able to find ways of calculating the bounds on the sigmoid curve with only a few operations of a desk calculator.

Mathematical convenience was still important in 1949, when Litchfield and Wilcoxon (1949) first discussed Bliss's probit model to describe the pattern of doses that kill animals in acute toxicology experiments. In fact, the major mathematical contribution of the Litchfield—Wilcoxon article was that it was able to replace Bliss's desk calculator procedures with graphical ones. To show that their graphical methods produced the same answer, Litchfield and Wilcoxon reproduced Bliss's formulas and showed some sample calculations. By one of those quirks of scientific tradition, the first computer programs that were written to implement the use of probits in acute toxicology used the formulas as they appeared in the Litchfield—Wilcoxon article. Since almost everyone who calculates an LD50 uses a computer program and since almost all such computer programs are descendents of the first few based on the Litchfield—Wilcoxon article, that article is cited as a reference in most published papers where an LD50 is determined. As a result, the 1949 Litchfield—Wilcoxon article is one of the most widely quoted in the biological literature. It is even cited for material that was peripheral to the purpose of the original authors, and poor Chester Bliss, who originated it all, is seldom given credit.

Notice that in Figure 5, which displays the best-fitting probit and the 90% confidence bounds on that curve for a given set of data, the width of the confidence bounds increases as the graph moves away from 50% probability. This can be seen by drawing horizontal lines at the 10, 50, and 90% points on the y axis until they intersect the bounding lines, then dropping vertical lines to the x axis, and reading off the logarithms of the doses associated with the confidence bounds. This has been done for the 90 and 50% probits in Figure 5. The reader should do the same for the 10% probit as an exercise in understanding this type of graph.

The above illustrates that for Bliss's probit the most precise estimate of a "dose that kills" can be made for the dose that

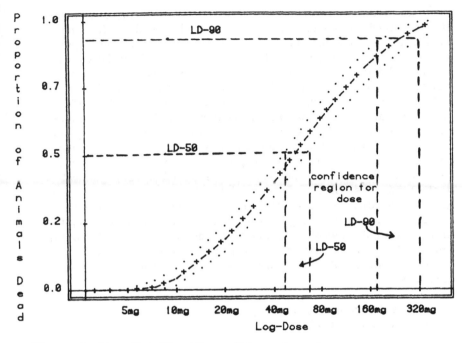

Figure 5 Sigmoidal fit with confidence bounds.

will kill 50% of the animals, the LD_{50}. Notice, in particular,
that attempts to estimate doses that kill 1 or 99% of the animals
will result in very wide confidence intervals, often so wide as
to be useless for any reasonable scientific investigation. This
is a property of all methods of investigating "doses that kill,"
as long as they are formulated in terms of random deadly doses
associated with individual animals, because the curve modeled
is a sigmoid, and any curves bounding the ends of the S-shaped
curves will tend to be nearly parallel. Thus any attempt to
run a horizontal line across the two bounding curves in order to
find the associated doses at very low or very high probits will
automatically lead to wide bounds. Furthermore, if the x axis
is in terms of the logarithm of the dose, then the actual doses
associated with the bounds will differ by orders of magnitude.

Since Bliss's probit curve is determined by only two parameters, the LD_{50} and the slope of the linear portion of the
curve, one can expect that it will not "fit" well to all possible
data. So the 1949 Litchfield–Wilcoxon paper includes a test of
goodness of fit. This is a standard component of any statistical

analysis of data. In order to analyze a set of data, the scientist is forced to choose a mathematical model, which will constrain him to consider only a small subset of the possible patterns of underlying probability. Thus any good statistical analysis will include a test of whether the data are really appropriate to that model. The theory of statistical tests will be discussed in a later chapter, but the user of a "black box" which calculates an LD50 and associated bounds should understand how to use the computer output that discusses the goodness of fit. In some computer outputs it will be called a test for *goodness of fit*; in other outputs it will be called a test for "linearity." Litchfield and Wilcoxon proposed a very general test of goodness of fit. Later articles (and some computer programs) use different tests that are proposed to be better in some sense. At this stage the student need not worry about this type of discussion. It should be enough to be able to find the goodness-of-fit test in the computer output and use it.

The goodness-of-fit test will usually be in the form of two numbers:

1. *The test statistic* (χ^2, F, t)
2. *A significance level*

They are equivalent; one number gives exactly the same information as the other. Let us concentrate, then, on the significance level (sometimes labeled p= or *p value*). The idea here is a very simple one (although it is bound up in a traditional terminology that carries implications from other scientific fields and often leads to confusion.) We assume that the mathematical model being used is correct, and we calculate the probability of the data falling the way they did. If the observed data are highly improbable under the model, then we can come to one of two conclusions: Either we have observed something that is highly improbable or the mathematical model is incorrect. We do not often observe improbable events (by definition), so we use a small p value as "evidence" against the truth of the mathematical model.

How small must the significance level, or p value, be to be called highly improbable? There is no "correct" answer to this question; it is a matter of literary taste. R. A. Fisher, the founder of modern mathematical statistics, used to consider three ranges of p values. If the p value was less than .001, then he took this as adequate evidence that the model was incorrect. If the p value was greater than .20, then he accepted the mathematical model for the moment, barring any other evi-

dence against it.   For intermediate p values he discussed how
one might design another experiment to elucidate the true mathematical model.   Fisher is also responsible for another version
of the "correct" p values.   For one of his elementary textbooks
(*Statistical Methods for Research Workers*; Fisher, 1925–1956)
he tabulated the p values associated with certain test statistics
(the other number printed out in most computer outputs).
Since the full table of values he wanted to use was already
copyrighted by a rival statistical group, he published only the
values associated with significance levels of 5 and 1%.   This
took place in the 1920s, but so pervasive has been the influence
of Fisher's work that p values of 5 and 1% are now widely accepted in the scientific world as "evidence of significance."
There are alternative approaches to this question, which will be
discussed in a later chapter.

   What should the toxicologist do when the goodness-of-fit test
produces a highly improbable significance level?   Alternative
mathematical models are discussed in the following section.

Alternative Mathematical Models

Sometimes the data do not fit Bliss's probit model at all well.
This may be because they are derived from an experiment
which, for theoretical reasons, should produce a pattern of
deadly doses that follows a different probability law.   It is not
often, however, that enough is known about the underlying
mechanism of toxicity to be able to calculate a reasonable probability law that might be expected.   More often the toxicologist
is in the position of someone who has a large number of keys
and must find the one that opens the lock.   Thus a large collection of mathematical models has evolved which can be applied
to the data of an experiment.   Recall that Bliss's probit model
has three characteristics:

1. There is no way of representing a dose associated with 0 or
   100% mortality.
2. The curve drops off sharply toward 0% and 100% outside the
   linear portion.
3. The curve is entirely determined by two parameters.

   We can get other models by changing one or more of these
characteristics.   We can keep the second and third ones but
account for doses with 0 and 100% mortality by using the *angle
transform*.   Not only does the angle transform allow one to find

doses associated with 0 and 100% mortality, but condition (2) is accentuated and the sigmoid drops off more rapidly outside the linear portion than the probit sigmoid. As with the probit, the curve is entirely determined by two parameters.

A model that retains conditions (1) and (3) but provides a curve with more gradual slopes to the ends of the *S* is the *logit curve*

The probit, angle transform, and logit curves are the most widely used sigmoids in toxicology. Figure 6 displays the three curves with the same parameters (the LD$_{50}$ and the slope of the linear portion) set at the same values so that the reader can see how they differ at the extremes of the sigmoid. It has been the practical experience of most applied statisticians that all three models lead to approximately the same results, as long as one is interested in the LD$_{50}$ and its bounds. The models differ considerably, however, if one attempts to calculate doses associated with very small and very large probabilities.

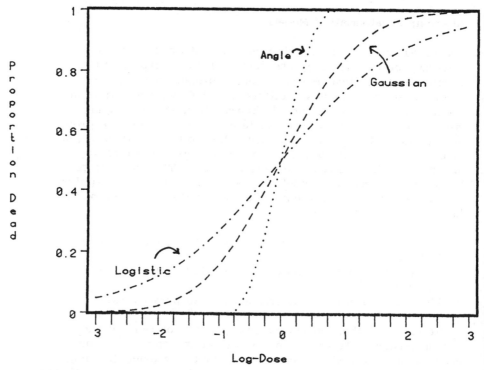

Figure 6 Three different sigmoids with some slope.

There are also model-free approaches to the problem. If the toxicologist is interested only in the LD$_{50}$ and its bounds, then it does not matter how the mathematical model treats the tails of the sigmoid. It is sufficient to know that some doses are below the LD50 and some above. The *Spearman–Karber* model (if one can call it a model) computes the probability of death at each dose tested, ignoring the other doses, and then combines them to produce an estimate of the LD50 and the bounds on it.

Greek legend tells of a bandit named Procustus who guarded a mountain pass and stopped all travelers on their way. He had a bed and would force the traveler to lie on it. If the traveler was too tall for the bed, Procustus cut off his legs; if the traveler was too short for the bed, Procustus stretched the traveler out, pulling out his legs and arms till he fit. The sigmoid models, whether they be probits, angles, or logits, are Procustian beds. They do not allow for dose reversals. They require that the pattern of mortality follow a constrained two-parameter class of curves. The Spearman–Karber method does no such thing: It allows for apparent dose reversals, or at least it accommodates them; it does not care about the relationship of the slope of the linear portion of the curve to the ends of the sigmoid. But, to gain this degree of flexibility, it gives up two things. It cannot be used to characterize the dose response, except at one point—the LD$_{50}$. Second, it produces wider confidence bounds.

The wider confidence bounds produced by the Spearman–Karber method make sense if one thinks about the problem. When the data are fit to the probit curve, the confidence bounds represent the degree of uncertainty associated with the fact that there might be other values of the two parameters of the probit that could fit the data, but only other probits are considered. When one does not impose a particular model (as with the Spearman–Karber method), then the confidence bounds have to represent a much larger class of sigmoids that might fit the data.

There is still another model-free method for determining the LD$_{50}$. This method requires a change in the design of the experiment. The usual LD$_{50}$ experiment calls for a fixed number of animals (usually 5–10) to be administered each of a predetermined set of doses and the resulting mortality is fed to a computer program. Another experimental design, due to Dixon and Massey (1969) and originally developed to determine the potency of different lots of explosive, requires that the toxicologist use one animal at a time. A set of doses is selected in advance to cover the range of what might be expected to be the

LD_{50}. For example, the doses might be 0.05, 0.5, 5, and 50 mg, or, to get a finer-tuned estimate, they might be 0.01, 0.02, 0.04, 0.08, 0.16, 0.32, 0.64 mg, and so on. A starting dose is chosen arbitrarily (usually somewhere in the middle). If the first animal dies at that dose, the next lowest dose is used for the next animal. If the first animal lives at the starting dose, the next highest dose is used for the next animal. The experiment continues sequentially: The dose is increased one step if the previous animal lived and decreased one step if the previous animal died. After a certain number of dose reversals (determined by the accuracy desired by the toxicologist) the experiment stops. The estimated LD_{50} is the average of all doses after the first reversal (including the "next dose" that would have been used had the experiment not been stopped). Confidence bounds are based upon the range of doses needed between reversals.

There are other, more complicated wrinkles among the possible methods of estimating LD_{50} values and plotting sigmoids. These include multiple up-and-down methods (where, instead of one animal, three, five, or seven are exposed to each dose), three- and four-parameter sigmoids, LD_{50} estimates based upon truncating sigmoid curves at predetermined points, "robust" techniques that allow for Procustian beds with somewhat adjustable head- and footboards, and alternative definitions of *best fitting*. At this point the author falls back upon the disclaimer found on most over-the-counter medications—if cough persists, consult a doctor: If the toxicologist is interested in such wrinkles, he should consult his friendly nearby biostatistician.

What Is Wrong with It All?

Figure 7 displays the results of a typical LD_{50} experiment for a compound never before investigated. In this study the technician had no idea of what dose would have any effect. She took one mouse and injected it with 0.5 mg of the compound with no apparent effect, and then tried three mice at 5 mg. All three died within an hour. This gave her a range of doses with which to try a formal experiment:

0.5 mg
0.7 mg
1.0 mg
1.5 mg
2.0 mg
4.0 mg
5.0 mg

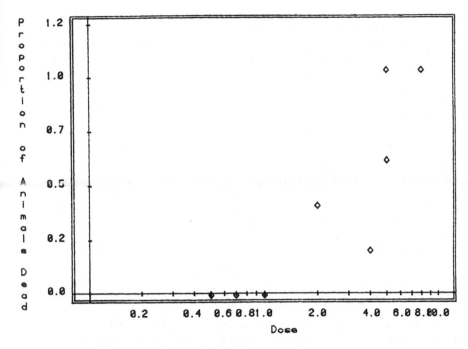

Figure 7 Results of a typical LD50 experiment.

Since Bliss's model (and most other sigmoids) use the logarithm as a dose metameter, a doubling of dose represents a unit difference on the logarithm scale. So the technician built her proposed doses around a doubling pattern (0.5, 1.0, 2.0, and 4.0 mg) but threw in some intermediate values. (Did she not trust the conventional wisdom that claims that incremental effects are seen only with a doubling of dose, which results in the use of log−dose?)

Using 5 mice per dose, the technician obtained the following data:

Dose (mg)	Number dead per number exposed
0.5	0/5 (0/1 from first trial)
0.7	0/5
1.0	0/5
2.0	2/5
4.0	1/5
5.0	3/5 (3/3 from first trial)

Failing to reproduce her 100% mortality at 5.0 mg, she tried
five more mice at 8.0 mg and killed five out of five.

This experiment is not like the theoretical "trial" presupposed
by the mathematical derivation of Bliss's probit. There are several doses that are too low to elicit a "response." The three
most widely used sigmoid curves (the probit, the angle, and
the logit) are symmetrical curves, requiring that the pattern
on the upper part of the S be the same as that on the lower
part of the S. So the inclusion of the three 0% responses will
force the curve to ignore the less than 100% response at 5.0
mg and force it to look more like a 100% response at 8.0 mg.
Litchfield and Wilcoxon recommend that the toxicologist use no
more than two consecutive doses of 0 or 100% response, so they
would recommend that the lowest dose be dropped, but they offer no advice when there is a skew in the data and show no examples of such a problem.

What about the 8.0-mg dose? Should it be used? Theoretically, the estimation of the LD50 and its bounds is based upon
mathematical theorems that assume that the choice of doses was
made independently of the events observed. If we use only the
doses up to 5.0 mg, the estimated LD50 is greater than 5.0 mg.
If we include the fact that 100% of the mice died at 8.0 mg, the
LD50 comes out below 5.0 mg. If we use the 5 out of 5 mortality at 8.0 mg, why can we not use the 3 out of 3 of mortality
in the initial test of 5.0 mg? After all, the initial choice of
5.0 mg was not based upon observations of random data.

The experiment described above may seem unusually sloppy.
It does not follow the guidelines for LD50 ingestion studies pro-
mulgated by the U. S. Food and Drug Administration; however, it
is typical of the sort of initial examination of a new substance
that occurs in a research environment. It also points up a fact
that is ignored in the theoretical development of statistical methods
for the analysis of LD50 experiments: The experimental design
and the events that occur during the experiment are often far
more important than the choice of mathematical model when it comes
to the final calculations. There are often questions concerning the
applicability of the mathematical model that are a fundamental part
of the applied science of toxicology. These questions can be effectively addressed only if the toxicologist who ran the experiment
and knows what actually happened retains control of the analysis
of the data.

Ideally, there is a consulting statistician who can aid the
toxicologist, but the resulting collaboration will work best if the
statistician is fully aware of the details of the experiment as it
was actually run and if the toxicologist is alert to the possibility

that the analysis might produce biological nonsense. The LD_{50} and its bounds are supposed to be estimates of the doses that will kill half the animals that might be drawn from a large pool of the same species and strain. The toxicologist should be satisfied that the final calculations make "sense" by answering the following questions:

1. Did approximately half the deaths occur at doses above the best-fitting estimate of the LD_{50}?
2. How wide are the bounds? If they are so narrow that they do not include doses that might be reasonable candidates for the LD_{50} under slightly different circumstances, then the mathematical model may be too restrictive.
3. How wide are the bounds? If they are so wide that they include doses that do not make biological sense, then the experiment may have been too small or might have some problems of design or the mathematical model may be assuming something about the dose—response curve that is not true.

The third item above needs some additional explanation. Requiring that the results make biological sense means more than looking at the numbers to get some sort of "gut feeling." This is what the statistical analysis did. It looked only at the numbers from the experiment. However, no toxicological study is done is splendid isolation. There are other studies run on the same or related compounds. Something is usually known about relative potency and expected modes of toxicity. A good experimental scientist begins an experiment with some idea of what the results should look like. An experimental result greatly at variance with what is expected means that either the experiment was badly done, the statistical analysis is inappropriate, or the scientist made an unexpected and highly significant breakthrough. Unexpected, highly significant breakthroughs occur very seldom.

Items (2) and (3) above also refer to the inappropriateness of the mathematical model. In some sense, all mathematical models are inappropriate. All mathematical models are attempts to impose a degree of order that can be comprehended by the finite human mind on the immense complexity of biological activity; they can never accurately describe reality. So we are constantly compromising. We create Procustian beds to fit our observed data and must always be alert to the possibility that our conclusions are based more on the length of the bed (the arbitrary assumptions of the mathematical model) than on what the data say.

2

Lessons from the LD$_{50}$

Overview

What is the dose that kills? The question put this way is ill
formed, since it is obvious that a dose that kills a given animal
may not kill another of the same species, strain, and sex. In
order to apply mathematical reasoning to the question, it has to
be rephrased and thus requires the construction of an abstract
mathematical framework within which it can be rephrased. As
anyone who remembers the "word problems" of high school algebra can testify, this is the hardest part of mathematics. When
we are faced with an entirely new question, the formulation of a
mathematical framework takes a leap of imagination bordering on
genius. Thus Chester Bliss, faced with the question of what
dose of an insecticide kills, set the pattern we have since followed, by conceiving of each insect as carrying a tag that gave
it a unique lethal dose. Bliss rephrased the question as one
that examined the probability distribution of these tags.

Bliss's formulation is not the only one that could have been made. We could have rephrased the question as "What dose is sure to kill?" and sought a dose that killed every animal on which it was tried. We could have asked, "What dose will not kill?" and sought a dose low enough to be safe. (The search for "safe" doses will be discussed in Unit II.) We could have viewed the question as a problem in pharmacology and biochemistry and sought to trace the metabolic fate of the administered compound, grinding up slices of tissue to determine the ultimate toxin in the target organ and constructing exquisite in virtro studies of receptor sites.

The point here is that there is no "correct" mathematical formulation when a question is as vague and ill formed as this one. Any mathematical formulation, however, will dictate the way in which we view the problem. It will also dictate the design of the experimental trials that will be used to answer the rephrased question. Let us see how Bliss's formulation dictates our view of this question and how we design trials.

Conceptually, we are replacing the complex organism of a living creature by a single number, the tag that tells us that animal's lethal dose. We act as if each animal comes with its own fixed and immutable tag, and the pattern of these tags becomes the question to be investigated. We view an experiment in which a single animal is given a predetermined dose of toxin as an indirect way of obtaining some information on that animal's tag. The information we get is quite minimal. After the test we know only whether the tag is a number above or below the predetermined dose that was given. The questions looked into then become abstract investigations about the distribution of tags as they can be determined by such minimal information.

Thus the first things the standard LD50 formulation shows us are the following:

1. Vague questions need to be replaced by well-defined mathematical formulations.
2. There is no single "correct" formulation, but we usually develop a "standard" formulation that is widely accepted in the scientific world.
3. The "standard" formulation dictates both the way in which the question is now posed and the type of experimental trial that is used to answer the question.

Consequences of Bliss's Formulation

Once we accept Bliss's formulation, we have to conclude that there is no such thing as a single number representing "the dose that kills." If we seek a single number, it has to do with the "average" behavior of a typical animal, a central tendency among the tags we conceptually hang on each animal. An easy choice of such a number is the median dose, the dose associated with a 50% probability of death, the LD_{50}, but it is not the only possible choice. We could have chosen the dose associated with a 10% probability of death, the LD_{10}, or one associated with some complicated function of the distribution of lethal doses. (We shall see such functions when we consider Sielkin's "mean tumor free time" in Unit III.)

In some sense, the LD_{50} is a poor answer to the initial question. What typical "man in the street" would feel safe if he knew only the dose of a compound that has a 50% probability of death? He would ask, "Is half the LD_{50} safe? Is 1/10 the LD_{50} safe?" But, as we saw in Chapter 1, the LD_{50} has some mathematical properties that make it a sensible number to seek. It can be found in the linear ascending portion of the sigmoid, so confidence bounds on the sigmoid lead to narrow confidence intervals on the LD_{50} (narrower, that is, than confidence intervals on doses associated with the tails of the sigmoid.) Second, the estimation of the LD_{50} and its bounds seems to produce similar numbers, regardless of the mathematical model (probit, angle, or logit) used.

Thus the mathematical formulation leads us to a question (and an answer) that is somewhat removed from the original question. It may not be particularily useful to the scientific problem at hand, but it is dictated by the nature of the mathematical formulation. We have a relatively precise answer to a well-defined question rather than what might have been a useful answer to our original, vague question.

Another consequence of Bliss's formulation is that we usually fit the data to a small class of sigmoids (such as the probits). If we attempt to interpret the best-fitting sigmoid that results, we have to be wary that our interpretation will be influenced less by the data than by the arbitrary assumptions of the model we chose. As long as we examine the LD_{50} and its neighbors, this is usually true. However, once we start looking at the tails of the distribution, we may get into situations where the arbitrary assumptions about the structure of these tails are more important than the data.

We have now added two more general lessons from examining the LD50:

4. The mathematical formulation modifies the original ill-defined question and produces a well-defined question and an answer of known precision, but the new question and answer may be somewhat removed from the original scientific problem.
5. The mathematical model may impose arbitrary assumptions that will influence some of our conclusions more than the data.

Comparative LD50 Values

One of the uses for the LD50 is to compare the toxic potentials of different compounds. Since we have a single number to represent "the dose that kills," we can compare it across similar compounds.

Thus, if compound A has an LD50 that is twice that of compound B, we can claim that B is twice as potent as A. But, as we saw in Chapter 1, the LD50 is inadequate to describe either the slope of the linear portion of the sigmoid or the behavior in the tails, and it becomes necessary to first see whether the sigmoids for the two compounds compared are parallel. This leads to the following:

6. The mathematical model can often be used to get a simple one-number answer to a complicated question, but it may be too simple an answer.

Some Comments on Confidence Bounds on the LD50

In Chapter 1 we noted that the best-fitting sigmoid curve can be bounded above and below by other curves such that we are reasonably sure that the "true" sigmoid lies between them. We have avoided the question of what *reasonably sure* means and how one goes about computing the parameters of those curves. The methods of computation will be postponed to Chapter 3, but it is necessary at this point to grapple with the phrase *reasonably sure* or the (apparently more precise) phrase *90% confidence bounds*.

There are several philosophical positions one can take concerning the meaning of confidence bounds, and these depend upon how one defines probability. For the experimental scientist, however, there is one way of viewing confidence bounds that would be accepted by protagonists of all of these philosophical positions. We conceive of the experiment which has just been run as one of a very large number of feasible experiments of the same design. The best-fitting sigmoid that emerges from this experiment will obviously differ (if only slightly) from the best-fitting sigmoid from another experiment of the same design, because each experiment is relatively small and uses different animals (with different tags). We construct the 90% confidence bounds so that if we were to run this very large number of experiments, 90% of the time or more the best-fitting sigmoid would lie between these bounds.

How do we know the best-fitting sigmoid will lie between the bounds 90% of the time? We "know" this because we assume that the narrow class of mathematical models consists of true descriptors of the distribution of tags for all the animals available to us. If we chose the wrong class of mathematical models, then the confidence bounds would be wrong. Since we can never know what the true underlying mathematic model is, we can never know whether we are right or wrong. In fact, we can be fairly confident that we are wrong. The usual models applied (the probit, the angle, and the logit) are relatively simple mathematical structures with only two free parameters. As we probe deeper into the complexities of biology, it becomes clearer that such simple two-parameter models are inadequate for describing reality.

The biostatistician will respond to the previous paragraph by stating that most of the procedures we use (such as calculating an LD50 based upon a probit model) are "robust" to slight errors in the initial mathematical formulation. If the bounds will not contain exactly 90% of the best-fitting curves, then they will contain something close to 90%. But do not ask what *close to* means in the previous sentence. Thus we have replaced an originally vague and ill-defined question with a fairly precise question and an answer whose relationship to reality is as vague as the original question.

For the experimental scientist, however, all is not lost. The principle of biological replication can rescue us. If the LD50 and its bounds mean anything, then we can not only conceive of a large number of similar trials, but we can also actually run other similar trials at different times in different labs. If we obtain estimates of LD50 that are close together and, more importantly, if the bounds all overlap, so that all the experiments

predict a similar range of best-fitting curves, then we can be
confident that the hunt for the LD$_{50}$ is a biologically consistent
procedure and it can be assumed to have some meaning and use.

The widespread use of computer programs based upon the
1949 Litchfield—Wilcoxon paper enabled Weil and Wright (1967)
to assemble a large number of published LD$_{50}$ trials on the
same compounds to see if the 90% confidence intervals did, in
fact, overlap. For most of the toxins examined, the confidence
intervals did not overlap; in fact, they differed so greatly as
to cast serious doubt on the validity of the whole procedure.

Why did this happen? Was it because the specific mathematical models (the probit, the angle, or the logit) imposed too
high a degree of arbitrary structure and masked what the data
said? Was it because Bliss's original formulation was inappropriate to the problem at hand?

Computer studies of robustness, where the computer gen-
erates data following a more complicated pattern of distributions
and the data are then fit to probits or logits, indicate that the
LD$_{50}$ and its confidence bounds are estimated with a high degree
of accuracy with these simpler, arbitrary models even when the
true model is more complicated. So it is more likely that Bliss's
original formulation is at fault. But it is not completely at
fault. It is a useful formulation, but its adaptation to studies
involving whole mammals tends to ignore another basic statistical idea, the concept of components of variance.

Bliss's original formulation dealt with insects, which have a
brief and brutish life, often of only a few days' duration, seldom for an entire season. From our macroscopic point of view,
insects are so numerous and identical in structure that they are
hardly more than simple tags with specific lethal doses marked
on them. Whether we examine several thousand mosquitos of a
particular strain today or tommorow or next week, the pattern
of tags will tend to be similar. At least, that was Bliss's experience, and his LD$_{50}$ values were consistent from trial to
trial.

But a rodent is more complex and subject to a greater degree
of genetic diversity. The sensitivity of a rat to a particular
oral toxin will depend upon the food in its stomach, the humid-
ity in the lab, the handling of the technician, its age, and per-
haps even the phase of the moon. In a single experiment a particular technician will draw 50—60 animals from a particular group
of newly arrived boxes, in a particular lab, over a brief interval of time. The "tags" that result on these animals do not represent a purely random selection of all possible "tags." And so,

the distribution of lethal doses we impute to the experiment as a result of its data will be a distribution of lethal doses only for *that* lab, under *those* conditions, for *that* technician, and at *that* time.

There is a traditional statistical model to account for a situation like this. We conceive of the lethal dose for a particular rodent as being a function of the lab, the time of year, specific aspects of the animal's condition, and so on. Since many of these factors are not easily determined or measured, we can conceive of a nesting of effects. We think of a very large operation, where a number of labs are running LD$_{50}$ tests at the same time. Thus the largest outer division is the time of year. At any particular time of year there are different labs; each lab there are a number of different boxes of rodents; each box of rodents is assigned to a small number of different technicians; each technician runs an independent LD$_{50}$ experiment.

If the resulting estimates of LD$_{50}$ have the same overall average value for each of the components of the nesting, then we have a *components-of-variance* model; that is, we assume that the underlying average LD$_{50}$ is the same regardless of time of year, lab, box, or technician. Then each element of the nesting contributes an additional component of uncertainty or variability. The overall variance (a specific mathematical measure of the variability) can be written as the sum of four variances: the variance due to time of year, the variance due to lab, the variance due to box, and the variance due to technician.

The width of the confidence bounds on the LD$_{50}$ from a given experiment are based upon the variability or scatter of data observed in that experiment. In the components-of-variance model the observed variability within a given experiment is measured entirely by the variance due to technician. If the other components of variance are negligible, then the computed confidence intervals will be wide enough to include most of what might be expected, and the best-fitting sigmoids from most of the experiments will tend to fall within the bounds computed from any one of them. If, however, there are positive components of variance due to box, lab, or time of year, then the computed bounds will be too narrow.

There is one way to determine if the components of variance due to time of year, lab, or box are negligible, and that is to run a large experiment in which the same toxin is examined at different times of year, in different labs, with different boxes, and with different technicians. In some sense, Weil and Wright's examination of published data was an experiment of this type.

They did not know about times of year or boxes of animals, but they were able to gauge the components of variance due to technician and lab from the published confidence bounds. The fact that the confidence bounds failed to overlap showed that the component of variance due to labs was far from negligible.

Thus a final lesson to be learned from examining the LD50 is the following:

7. The basic mathematical formulation may assume a degree of homogeneity to the randomness that leads to confidence bounds which are too narrow to represent reality. There is no substitute for biological replication in another lab at another time as a check on this possibility.

3

Theory Behind the LD$_{50}$

Concept of a Probability Distribution

Statistical theory concerns itself with patterns of numbers.
Suppose we had a set of 1000 numbers that resulted from rolling
a single die. We might have 144 occurrences of the number 1,
180 occurrences of the number 2, 181 occurrences of the number 2, 181 occurrences of the number 3, and so on. If we
plotted the frequency of a given number, we would obtain a
histogram, as in Figure 1. If we were to roll that single die
a million times or 10 million times, we could still plot the percentage of times we observed a 1 or a 2 or a 5, and so on. If
the die were perfectly balanced, we would expect, in the long
run, to obtain the same percentage of 1's as of 2's or of 5's
and so forth. One-sixth of the time the die will tend to come
up with a given number, and the proportion of times we obtain
a given number could be plotted as a histogram as shown in
Figure 2.

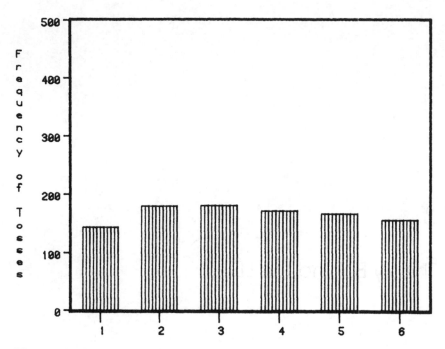

Figure 1   Frequencies in 1000 tosses.

Frequency plots such as those in Figures 1 and 2 describe
a pattern of numbers completely, provided that only a finite
collection of numbers is possible. (For the die we can get only
1,2,3,4,5, or 6, regardless of how many times we throw the die.)

Modern statistical theory can be said to have originated with
the founding of the journal *Biometrika* in 1898 by Francis Galton, Karl Pearson, and Raphael Weldon. Their stated purpose
was to examine the nature of change that could be observed in
living organisms which might occur due to evolution. They recognized that the grossly observable aspects of evolution, such
as the rising of a new species, takes a period of time far beyond
the range of human history, but they thought they could dis-
cern more subtle events by observing the changes that occur
in the frequencies associated with specific measurements. For
example, correspondents for *Biometrika* raided ancient Egyptian
graves and poured shot into the skulls they found to measure
cranial capacities; these were then compared to the cranial capacities measured in modern Egyptian skulls. They collected
crabs from the silted areas of Plymouth harbor and compared
the widths of their shells with those from crabs collected from
unsilted areas.

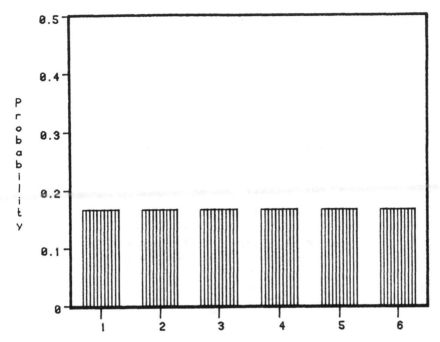

Figure 2 Theoretical frequencies.

The idea was that one might not be able to see gross changes in average measurements over short periods of time, but the probability distributions of some particular measurement might show a shift with time that reflected the increasing probability of survival for organisms with some minor structural advantage. The concept of a probability distribution has undergone subtle changes since then, and there is a raging philosophical debate about the meaning of "probability" when it is applied to scientific observations. However, for the experimental biologist (like the toxicologist), the original formulation due to Galton et al. is perfectly adequate, and we will plunge no further into that morass than they did.

The Galton—Pearson—Weldon formulation assumes that we can measure something about each animal and obtain such measurements from a very large collection of similar animals. If it were possible to measure this thing on all these animals, we could then construct a frequency histogram in a similar way as we constructed Figure 1. In some cases it might be possible to derive a theoretical distribution based upon examining how the measurement came about, as we did for the single die in constructing

Table 1 Weights of Male Rats (in Groups of Five)

268	276	245	252	290	226
226	264	241	289	280	296
286	228	296	264	228	266
212	245	271	236	250	296
216	202	298	216	252	273
286	276	206	246	275	
220	280	256	272	264	
201	252	206	240	226	
219	231	232	225	245	
236	286	252	222	—	

Frequencies of Final Digits:

Digit	Frequency	Digit	Frequency
0	6	5	5
1	4	6	20
2	9	7	0
3	1	8	4
4	3	9	2

Number of Occurrences of "6" in Groups of Five:

0 times 1		3 times 1	
1 times 3		4 times 1	
2 times 4		5 times 0	

Figure 2. In cases where we could not measure all possible animals, we might be able to combine the measurements we could make with theoretical considerations and derive a distribution that fits the data.

Table 1 displays the weights of 54 male rats as determined by a single technician attending to a chronic toxicity study. The technician lifted each rat out of its cage, set it on a scale, waited for it to settle down, read and recorded the number on a tally sheet, and then replaced the rat in its cage. Since the animals were not always perfectly still and the entire operation was done very quickly, there was a certain amount of choice in deciding exactly what number to record.

When it comes to recording numbers, there is a psychological phenomenon known as digit preference. When a person records a number and knows that the final digit is not really important, his unconscious impulses take over and he tends to avoid recording those numbers his subconcious does not like and to replace them with numbers his subconscious likes. It was important to this technician whether the animals weighed 230 or 240 g, since such a difference would dictate a difference in feed calculations, but a difference of less than 5 g would be considered unimportant. A quick glance at Table 1 will show that this technician unconsciously preferred 6's and 2's. If there had been no preference, we would expect all 10 possible digits to show up with equal probability in the unit position. However, there are 20 numbers ending in 6 and only 5 ending in either 5 or 7; similarly, there are 9 numbers ending in 2 and only 5 ending in 1 or 3.

The actual operation of weighing required that the technician remove a rack of five cages, weight those animals, replace the rack, remove the next rack, and so on. Thus the numbers in Table 1 are organized in groups of five. These will be used to illustrate the Galton—Pearson—Weldon idea that one can combine observed data with theoretical considerations to derive a probability distribution that fits the data. This relatively simple structure is used so that the student can have all the pleasure and pain of going through the details of deriving a best-fitting distribution. For more complicated structures we will assume that the student has "black box" computer programs.

If the technician who recorded these weights was paying careful attention to the important part of his work, then the choice of the final digit for the weight of a specific animal would not have been influenced in any way by the choice of the final digit for the weight of the previous animal. This is an example of two events that are *statistically independent*. If the choice of a 6 over a 5 or 7 is really subconscious, then the probability that the technician will choose a 6 over the other digits, when there is an opportunity to choose, will remain the same, animal after animal.

In this way the final digits of the weights associated with a rack of five animals can be described theoretically as belonging to what is called a *multinomial distribution*. To keep things at their simplest, we will note only whether the final digit is a 6 or not. Then the number of times we get a final digit of 6 in a group of five weights will follow a reduced version of the multinomial distribution known as a *binomial distribution*. Table 1 also displays the number of times the final digit is a 6 for each of the 10 groups of five animals.

We do not know the subconscious probability of this technician preferring the digit 6, but if we assume it was always the

same, we can symbolize it by the letter p. Then the only way
for us to have all five numbers end in 6 is for five independent
events with probability p to have occurred, or the probability
of five out of five final digits being a 6 is p raised to the fifth
power. Similarly, the probability of not getting a 6 is 1 − p
and the probability of getting all five weights to fail to end in
6 is 1 − p raised to the fifth power. If we were to sit down
and enumerate all the possible patterns that might emerge along
with their associated probabilities, we would find that we can
construct a formula for the probability of getting exactly X
weights ending in 6. The formula would contain the symbol p
(whose value we do not know), but all other parts of that formula would consist of predetermined numbers.

These formulas are as follows:

p^5: all five ended in 6
$5p^4(1 − p)$: one of the five did not
$10p^3(1 − p)^2$: two of the five did not
$10p^2(1 − p)^3$: three of the five did not
$5p(1 − p)^4$: four of the five did not
$(1 − p)^5$: all five did not

If we knew the value of p, we could plot the frequency histogram, on a theoretical basis only. Since we do not know the
value of p, we estimate it from the observed data. We have a
number of choices in estimating the value of p: We could take
the frequency with which 6 occurred in the first five weights;
we could take the frequency of 6's in all 54 weights, we could
go back to the notebooks of this particular technician and count
how often his weights ended in 6 over the many years he has
been at this lab; or we could set up a psychological experiment
with the technician and try to determine the degree of his innate
preference for 6's.

These four possible procedures illustrate four general methods
for estimating the value of unknown parameters when trying to
fit data to a theoretical distribution. In the first procedure we
use less than all the data available from the experiment; in the
second one we use all the data from the experiment but ignore
any information we might have external to the experiment; in the
third we make use of prior information and assume that the underlying parameter has not changed; in the fourth we construct
an auxillary experiment to get a firm handle on the value of the
parameter. All four methods are perfectly "valid" and there are
conditions under which one might be preferred over another.

If we assume there is no other information forthcoming aside from what we have from the experiment and if we have no reason to want to save some of the data as a secondary check on the quality of the fit, then most statisticians will agree that the "best" method of estimating p is to use the overall frequency of 6's in the entire experiment. What makes one method of estimating better than another will be addressed later in this chapter.

Gaussian or Normal Distribution

On way to think of biological variability is to consider some underlying "typical" organism. For instance, we might conceive of a "typical" male Sprague-Dawley adult rat. All male Sprague-Dawley adult rats differ from this typical one because of an accumulation of slight differences in DNA structure or environment (both pre- and postnatal). In some sense, the weight of a given animal can be thought of as the weight of the "typical" animal plus the sum of a large collection of small numbers (both positive and negative), each one of them induced by some slight difference.

There is a deep and difficult ergoidic theorem in mathematics known as the *central limit theorem* which states that the distribution of random numbers that are created out of sums of large collections of very small, statistically independent numbers will tend to follow a certain theoretical pattern. Just as the theoretical binomial distribution had one unknown parameter, p, the distribution predicted by the central limit theory has two unknown parameters. The distribution predicted by the central limit theorem is called the *Gaussian* distribution (in honor of Karl F. Gauss, an early nineteenth-century mathematician who first suggested the central limit theorem in its full form). It is sometimes called the *normal* distribution or (more vaguely) the bell-shaped curve.

Because it describes the distribution of numbers that are sums of large collections of other numbers, the Gaussian distribution deals with an infinite set of possible numbers. That is, it is theoretically possible for any number, from $-\infty$ to $+\infty$, to occur when sampling from a Gaussian distribution. So there is no way to represent the frequency distribution as a histogram, since there is no way of representing every possible number as a bar of positive width. However, one can represent the "frequency" of a continuous distribution like the Gaussian as if the bars in the histogram had been replaced by dimension-

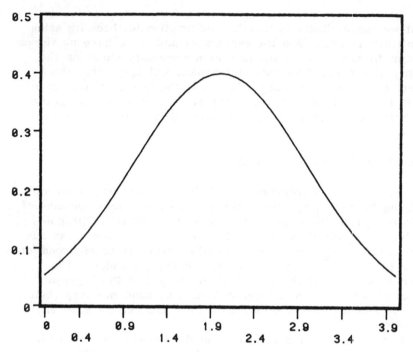

Figure 3   The Gaussian density function.

less lines, one above every possible number. If we strip away
the lines and keep only their topmost points, we get a curve
like that in Figure 3. To distinguish between this situation and
that in which only a finite set of numbers is possible (as in the
binomial distribution), such a graph is called a *density function*.
The probability of observing a number between two values is the
area under that curve from one value to the next. In Figure 4
the marked area is the probability of observing a number between 1.7 and 2.5.

    An alternative method of representing the Gaussian distribu-
tion is in terms of the *distribution function*. For each possible
value of x we plot the probability that we will observe a number
less than or equal to x. Such a plot is shown in Figure 5.
Since it is possible but highly improbable that we would observe
a number less than −1000, the curve starts on the left near
zero. It rises slowly until we get among numbers that are more
probable, at which point the curve rises steeply (almost as a

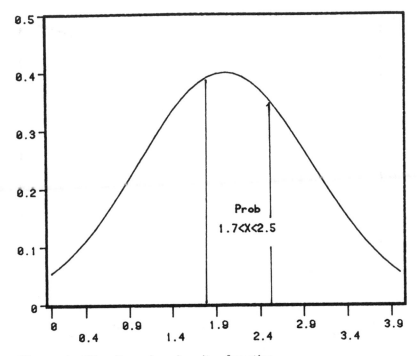

Figure 4   The Gaussian density function.

straight line) and then levels off, since the probability of observing a number less than 500 is close to 1.0 and not much different from observing a number less than 1000.   The student
will recognize the sigmoid of Chapter 1.

As noted above, there are two parameters that describe the
Gaussian distribution: One is the mean and the other is the
standard deviation.   In Figure 3 the mean is the value of x
over which the bell-shaped curve is at its peak.   The density
function curves downward on either side of the mean, or peak.
It continues to go down, but the curvature changes so that the
curve becomes concave upward instead of downward.   The point
where the curvature changes (the point of inflection) defines
the standard deviation.   The standard deviation is the distance
from the mean to the point of inflection, and since the curve
is perfectly symmetrical, this distance is the same, whether you
move to the left or to the right.

The two parameters can be seen (although less clearly) in
Figure 5, the distribution function.   The mean is the point as-

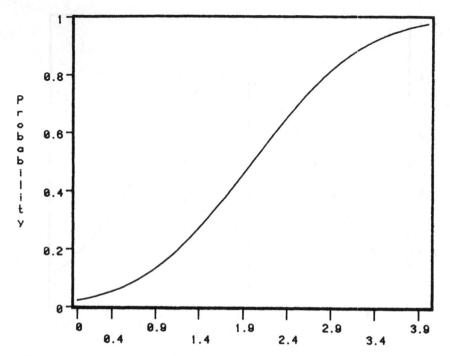

Figure 5 Gaussian distribution function.

sociated with the 50% probability. The slope of the linear portion of the curve defines the standard deviation.

The Gaussian density curve comes very close to 0 density
as it moves a short distance beyond the point of inflection: 95%
of the probability is contained within 1.645 standard deviations
of the mean, 99% is within 2.576 standard deviations of the
mean, and 99.9% is within 3 standard deviations of the mean.
A convenient concept is that of the *probable error*, which determines the middle 50% of the distribution. An observation
chosen at random is more probably inside this interval than outside it. Figure 6 displays the region of the probable error.

Beyond the Gaussian Distribution

In the last decade of the nineteenth century Karl Pearson proposed a modification of the central limit theorem which lead to
the *Pearson system of skew distributions* (Elderton and Johnson,

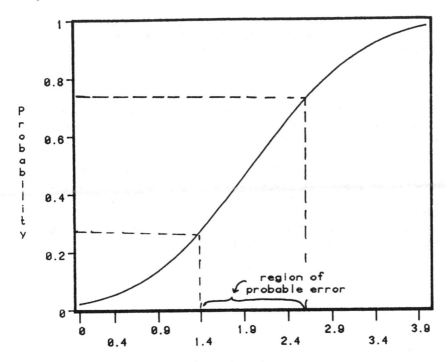

Figure 6 Gaussian distribution function.

1969). Until his death in 1935, many of the papers in *Biome-trika* were devoted to estimating parameters from these skew distributions when they were fit to observed data. Other classes of distributions that are not quite Gaussian have been proposed since then and many of Pearson's techniques for esti-mating parameters have fallen into disrepute, but the basic idea proposed by Pearson is still a good starting point to understand the development of non-Gaussian distributions.

Pearson's observations of biological data suggested to him that the things we observe (the cranial capacity of human skulls, the weights of rats, etc.) may result from the accumu-lation of a large number of deviations from the "typical" but that biological processes tend to distort and modify the final result. In some sense, we can think of an underlying random measurement that is the result of the sum of a large collection of slight deviations, but the final measurement we can make is the further result of a biological operation on that underlying measurement. If the underlying Gaussian measurement is denoted by a capital X, then we can think of the biological process as represented by some function that results in our observed measurement Y:

$$Y = f(X)$$

(Here we use the functional notation of Chapter 1, read "f of X," to show that something is going on but that we are not yet ready to display what that something is.)

One of the most useful theorems of calculus states that for a function that is smooth and continuous, that is, that does not have too many peculiar properties, it is possible to obtain good approximations by replacing that one function with a simpler, polynomial one in the argument.

Pearson used this theorem to replace the unknown function $f(X)$ with a third-degree polynomial:

$$f(X) = A + BX + CX^2 + DX^3$$

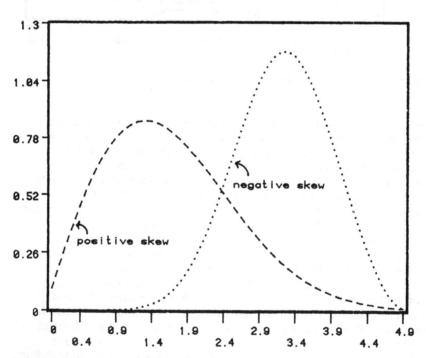

Figure 7   Skewed density functions.

In this way he obtained a system of distributions with four parameters. The parameters A and B correspond to the two parameters (mean and standard deviation) of the Gaussian distribution. The two new parameters correspond to the degree to which the new distribution differs from the Gaussian one.

One of the new parameters describes the *skewness* of the distribution. The bell-shaped curve of the Gaussian distribution is symmetrical, with probability patterns above and below the mean reflecting one another. Such symmetry is represented by a skewness coefficient of 0. Figure 7 displays density functions with positive and negative skewness.

The other new parameter Pearson called the *kurtosis*. This describes the rate at which a distribution or density function approaches the 0 line. Distributions with kurtosis greater than that of the Gaussian have tails in their sigmoid curves that go more gradually toward 0 or 100%. Figure 8 displays density functions with kurtosis greater and less than that of the Gaussian distribution.

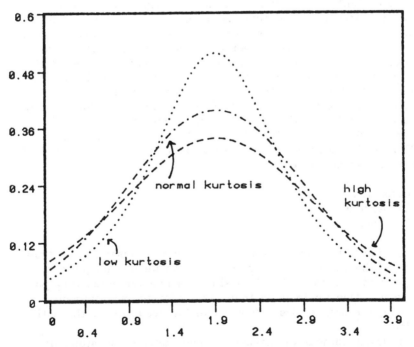

Figure 8 Density functions with the same variance but different kurtosis.

The essential idea in the Pearson system is that of the *parameter*. We assume that the numbers we can collect come from a distribution that belongs to some system of distributions (such as the Pearsonian system) which differ from one another only through the different values of their parameters. If we assume that the numbers come from a Gaussian distribution, we need only estimate the mean and the standard deviation to be able to specify the exact distribution for the data. If we assume they come from the larger class of Pearsonian distributions, we need to estimate the mean, the standard deviation, the skewness, and the kurtosis to find the exact distribution of the data.

In LD50 trials we usually assume that the data come from a family of distributions with two parameters and we attempt to estimate those two parameters.

Estimating Parameters

Because we can never obtain all the possible numbers that make up a probability distribution, we have to impute some sort of theoretical structure and then estimate the parameters of that small family of distributions from the data we can collect. This was illustrated in the first section of this chapter, where we deduced that the number of times a weight ending in the digit 6 in a group of five weighings would follow a binomial distribution. We then had to estimate only one parameter, p.

There is an elaborate theory of estimation in the formulation of mathematical statistics. Most of it is due to R. A. Fisher, where articles were published in the 1920s and 1930s (Fisher, 1925). Fisher identified three properties that a good method of estimation should have:

1. Consistency
2. Unbiasedness
3. Precision

Consistency means that we get closer and closer to the true value of the parameter being estimated as we use larger and larger sets of data. This is a sine qua non for estimating parameters. One would hope that, as a minimum requirement, increasing the sample size will improve the accuracy of the estimation.

Unbiased methods of estimation occur under a related but slightly different concept. We conceive of running a series of experiments of a fixed size over and over again. If our method of estimation is unbiased, then the average of the estimates from a very large number of similar trials will be very close to the true value of the parameter.

Consistency refers to what might happen in a single very large trial. Bias refers to what might happen on the average over a large number of small trials. It is possible for a method of estimation to be consistent but biased. In fact, there are times when other considerations make the "best" estimator one that is biased. As an example, it is common practice in municipal water systems to estimate the level of common bacteria (*Escherichiacoli*) in the water by a method known as the "most-probable-number" procedure. Various dilutions of the water in question are incubated so that the growth of bacteria may be observed. In order to make use of the data from tubes that show no growth, the only available method of statistical analysis requires the use of a biased estimator. All unbiased estimators for this situation will have a much greater probability of error.

And this leads us to the third criterion of Fisher's, *precision*. Precision refers to the width of a reasonable confidence bound on the underlying parameter. If the method of estimation is very precise, the confidence bounds will be narrow. If the method of estimation is less precise, the confidence bounds will be wider.

All methods of estimating a parameter require that we use available data. Since the data contain random noise, the resulting estimator is a random variable, and thus the estimator of a parameter has, itself, a probability distribution. Many standard methods of estimation lead to estimators whose distributions are Gaussian. In such cases precision can be determined by knowing the standard deviation of the estimator's distribution. The standard deviation of an estimator is usually called the *standard error of the estimate*

Fisher showed that there were many ways to estimate a parameter. The earlier example of the underlying subconscious probability that the lab technician will choose a 6 over a 5 or a 7 displayed four of these ways. Pearson had used a technique known as the *method of moments*, but Fisher was able to find others. Many of these methods led to unbiased and consistent estimates, so Fisher needed a way of determining a best one among them. He proposed that the "best" among them would be the one with the greatest precision. He then proved that if there is a method of estimation that is more precise than any other, it is the method of *maximum likelihood*.

In principle, the method of maximum likelihood is straightforward. We construct a mathematical formula that describes the probability of the pattern of data that actually occurred in terms of the unknown parameters. This is similar to the way in which one solves an algebra problem by setting a symbol for the unknown value in a mathematical description of the problem and solving for the unknown. However, Fisher's "solution" was to find those values of the parameter that maximized the probability of what had been observed. To avoid confusion between this mathematical formulation and the concept of probability as a measure of what has yet to happen, Fisher called the mathematical formulation the *likelihood* of the observations.

The major difficulty with this approach is that without a computer it is very difficult or even impossible to find the parameter values that maximize the likelihood for all but a few situations. Therefore, although Fisher was able to solve the problem of finding a "best" method of estimation in the late 1920s, the developments of statistical research that followed in the 1930s through the 1950s produced many ingenious methods of estimation that did not involve maximum likelihood, if only because the solution for a specific case was too difficult. The advent of modern electronic computers has put an end to this problem. It might be difficult or even impossible to find the maximizing parameter values with only the tools of algebra and calculus, but a computer can find them numerically through the use of iterative procedures that require millions of calculations. As a result, it is now possible to find Fisher's "best" estimates for any situation that can be written in terms of a family of probability distributions indexed by unknown parameters. The traditional uses of statistics in particular scientific fields, however, involve the earlier methods whose development was based upon the need to find analytic solutions through manipulations of algebra and calculus.

This is what has happened with the LD$_{50}$. The earliest attempts at finding the LD$_{50}$ (see, e.g., Burn, 1937) involved methods of analysis similar to Pearson's method of moments. When Bliss proposed the use of probits, he suggested a method now known as *linearized least squares*. With the aid of Fisher, Bliss then modified it and developed the *weighted linearized least-squares* procedure. The next section examines the difference between these methods.

Finding Consistent Estimators of the LD50

In order to estimate the LD50 from experimental data, we first
have to propose a general formula for the probability distribution
that lies behind the observed data. This formula or family of
distributions has at least one parameter, the LD50, that we seek
to estimate, and usually one other (which describes the slope of
the linear portion of the sigmoid). As indicated in Chapter 1,
three such families are usually used. There is no good reason
to prefer one or another nor is there any good reason to reject
one of the very large number of seldom-used alternative families
of distribution. So, for the moment, let us concentrate on the
general idea of finding the LD50 and use functional notation to
set aside consideration of the exact formula that will be used.
We then write

$$P(death) = f(dose; slope, LD50)$$

[Recall that the expression $f(\square)$ is read "f of" and that the expressions within the parentheses are arguments of the function.
This means that there is some sort of a mathematical formula
which includes a symbol for the dose that was used for a given
set of animals, as well as symbols for the two parameters of the
family of probability distributions, the LD50 and the slope of the
linear portion the sigmoid.]

Suppose we expose 10 animals to a dose of 5 mg/kg and 3 of
them die. We expose 10 more to 10 mg/kg and 8 die and we expose 10 more to 15 mg/kg and all 10 die. Ideally we seek to
solve the following set of equations:

$$3/10 = f(5; slope, LD50)$$

$$8/10 = f(10; slope, LD50)$$

$$10/10 = f(15; slope, LD50)$$

By some algebraic manipulation we might be able to find values
for the slope and LD50 that satisfy all three equations; however,
there are two unknowns and three equations, and chances are
that any set of values that might be found to fit two of the
equations will not fit the third. One way around this is to re-
write the three equations as

$$0 = 3/10 - f(5; slope, LD50)$$

$$0 = 8/10 - f(10;\ \text{slope},\ LD_{50})$$
$$0 = 10/10 - f(15;\ \text{slope},\ LD_{50})$$

We can now add up all the right-hand sides of these equations and find values of the slope and LD_{50} that make this sum closest to zero (the sum of the left-hand sides.) Such a procedure does not quite work, since the resulting estimates of the slope and LD_{50} will not be consistent; however, if we first square the right-hand sides and then sum the squares, the values of the slope and LD_{50} that minimize the sum of squares is consistent.

Each component of this procedure, that is, an expression like

$$x/n - f(\text{dose};\ \text{slope},\ LD_{50})$$

is called the *deviation of expected from observed*. The procedure of summing the squared deviations and finding values of the slope and LD_{50} that minimize this sum of squares is called the *least-sqaures solution*. This method of solving for best-fitting parameters goes back to the early part of the nineteenth century. It has the convenience of often allowing for a simple solution through calculus. Under certain circumstances (that do not occur in LD_{50} experiments) the solution produces estimates that are not only consistent but also unbiased and of maximum precision.

For an LD_{50} experiment, where the mathematical formula for $f(d;\ \text{slope},\ LD_{50})$ is based on the probit or where the logit is used, the least-squares solution is difficult to find, even with calculus. So Bliss proposed a modified version of least squares in his original formulation. He took the equation

$$P(\text{death}) = f(\text{dose};\ \text{slope},\ LD_{50})$$

and applied a modifying function to both sides until he had a formula of the form

$$M(P(\text{death})) = A + B(\text{dose})$$

where the left-hand side is read "M" of probability of death" and the right-hand side indicates that there is some fixed number A which is added to the product of the dose and a fixed number B. In this formulation B is the slope of the linear portion of

the sigmoid and A is related to the LD50. In the experiment
described above, we would then have three equations,

$$M(3/10) = A + 5B$$

$$M(8/10) = A + 10B$$

$$M(10/10) = A + 15B$$

or, rewritten,

$$0 = M(3/10) - (A + 5B)$$

$$0 = M(8/10) - (A + 10B)$$

$$0 = M(10/10) - (A + 15B)$$

The mathematical expressions for $M(x/n)$ can be very complicated but for the probit they had already been tabulated, so
Bliss needed only to look up the exact numbers associated with
$M(3/10)$, $M(8/10)$, and $M(10/10)$, put them into the formula for
the deviations of observed from expected, and find the least-
squares solutions by the standard, easily computed formulas of
linear least-squares that had been worked out by Laplace in the
early nineteenth century.

There were two problems with Bliss's "simple" solution. The
Gaussian probability distribution associates the value of $-\infty$ to
a probability of 0% and $+\infty$ to a probability of 100%. So, if one
were to observe 0 out of 10 animals killed at a given dose, one
would have to put $-\infty$ into the equations, or if one observed 10
of 10 animals killed, one would have to use $+\infty$. The existence
of such a symbol would immediately destroy the influence of the
finite numbers and the "least-squares" estimate of the LD50
would become either $+\infty$, $-\infty$, or indeterminate. In fact, if one
were to expose a large number of animals to a given dose (say,
100 units) and find very few or almost all of them killed, then
that observation would also dominate the "solution," pulling the
LD50 too far off. This demonstrates what it means to have a
biased estimate. Bliss's linearized least-squares solution requires
that observations of 0 and 100% be ignored and then produces a
consistent estimator that may be biased.

Consulting with Fisher, Bliss modified his original proposal
so that extreme observations would have less weight in the solution than those near the true LD50. He also developed a
technique for handling observations of 0 and 100% by first finding a solution that did not include them, estimating the probabil-

ity of death at the doses for which he observed 0 and 100%,
and finding a new solution with these estimates slightly adjusted
and used in place of the observed 0 and 100%. The result was
a consistent and unbiased estimate of the LD50 and the slope.
The original formulation by Bliss required a great deal of work
involving tabulated values and a desk calculator (remember, it
was designed for the technology of the late 1930s.) Litchfield
and Wilcoxon simplified these techniques by producing a set of
nomograms and showing how they could be used with log-normal
graph paper. An appendix to their article (Litchfield and Wilcoxon, 1949) replicates Bliss's formulas and demonstrates how
their graphical procedure is equivalent to his.

It is often possible to convert a problem to find a linearized
least-squares solution. This has the advantage of providing a
simple set of formulas for solution but the disadvantage of usually producing biased estimates. The bias can be overcome by
weighting some observations more than others. However, the
weighted linearized least-squares solution is often as hard to
calculate as the original least squares might have been. At any
rate, modern computers have fast accurate algorithms that allow
one to calculate the original least-squares solutions.

Fisher's maximum likelihood general solution for all statistical
problems can be introduced into the estimation of the LD50 in
the following fashion. We start as before with the functional
equation

$$P(death) = f(dose;\ slope,\ LD_{50})$$

Then, if we observe 3 out of 10 animals dead at 5 mg/kg, we
can write the probability of that observation (following the binomial distribution theory described earlier) as

$$120\ f(5;\ slope,\ LD_{50})^3[1 - f(5;\ slope,\ LD_{50})]^7$$

We can write similar expressions for the observations of 8 out of
10 dead at 10 mg and 10 out of 10 dead at 15 mg. The probability of observing all three events is then the product of these
three probabilities. Following Fisher, we regard the observations as fixed by the experiment and we find those values of
slope and LD50 that maximize this overall probability. The results are estimates of the two parameters, the slope and the
LD50, that are consistent, have maximum precision, but which
need not be unbiased. However, it is possible to use calculus

to determine the degree of bias these estimates will have and to correct them to unbiased estimates.

The three procedures for calculating estimates of the LD50 described above—the least squares, the linearized least squares (with or without weighting), and the maximum likelihood—all assume that the toxicologist has access to a computer or to the formulas and a programmable calculator. There are a number of "quick and dirty" approximations that have been developed that enable one to get a reasonably accurate approximation of these estimates. One method is to plot the observed proportions of animals dead against the dose on log-normal graph paper and fit a straight line by eye. This approximates the unweighted linearized least-squares solution. If the LD50 is going to be used only for a general "ballpark" estimate of an appropriate starting dose for a chronic study, such a method is often adequate.

The Spearman—Karber Solution

Statistical theory allows a number of different ways of describing the same situation. Some examples of this were shown earlier, when it was pointed out in Chapter 1 that the p value and the test statistic printed out by most computer programs for the LD50 are equivalent and that one can consider either the frequency—density curve or the distribution function to describe a probability distribution. Sometimes it is possible to manipulate one of the alternate methods of describing the same thing to avoid many of the difficulties that arise when one attempts to "solve" a problem with a standard approach. Such was the method evolved in 1931 by Karber (Burns, 1937).

Instead of trying to formulate the probelm as one involving the estimation of parameters for a well-defined family of distributions (Bliss's approach), Karber suggested making use of the mathematical theorem that the area under the distribution curve (the sigmoid) is equal to the LD50. This requires knowing the true formula for the underlying distribution to obtain an exact solution, but Karber pointed out that the area under the distribution curve can be estimated in the same way as the parameters of a given distribution.

Figure 9 shows the following set of data for doses of digitalis given to groups of five frogs (taken from Burn, 1937): Superimposed on the data points (marked by crosses) is a theoretical sigmoid curve. To approximate the area under the sig-

Dose (cm^3/100 g)	Proportion killed
0.9	5/5
0.8	4/5
0.7	2/5
0.6	4/5
0.5	3/5
0.4	1/5
0.3	0/5

moid curve, we only have to connect the data points, forming a group of six trapezoids. The total area of these six trapezoids estimates the area under the sigmoid curve and hence provides an estimate of the LD_{50}. This method of estimation is consistent but it need not be unbiased and it is definitely

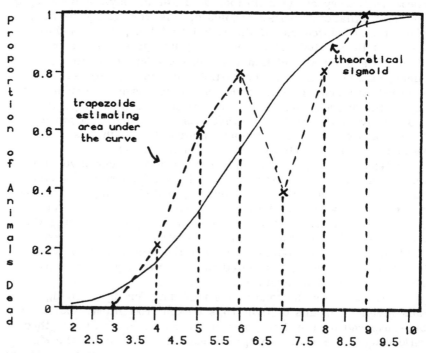

Figure 9 Spearman—Karber estimate of the LD_{50}.

not the most precise; however, it avoids imposing an arbitrary mathematical structure on the data.

There is one major problem with Karber's method (called the Spearman—Karber method to acknowledge prior publication of a similar procedure by the American psychologist Spearman.) If the frog experiment had used higher doses, one probably would have gotten five out of five kills at doses of 1.0, 1.1, 1.2, and 1.3 cm^3/100 g. This would have added three trapezoids (or, rather, rectangles) of area 0.1 units to the sum used in the calculation, increasing the estimate of the LD_{50}. A number of doses added at the lower end, on the other hand, would have added zeros to the sum and not affected the estimate. So Kar- ber proposed that the lowest dose that was reasonably sure of killing all the animals be taken and that all doses above that be ignored. Then the LD_{50} can be calculated in an inverse fashion by starting with this "sure kill" dose and subtracting the area under the observed curve. Karber also assumed that the proper metameter of dose is the log, and so the Spearman— Karber method requires using log doses in the calculation of the area under the curve.

Thus the formula for calculating a Spearman-Karber estimate of the LD_{50} is the following: Convert all doses to log doses, which gives the values $x_1, x_2, x_3, \ldots, x_n$. Let

p_i = (number of animals dead)/(number exposed), ith dose

Then log LD_{50} is calculated as

$$\log LD_{50} = x_n - \text{sum of } \tfrac{1}{2}(p_i + p_{i+1})(x_{i+1} - x_i)$$

where the sum runs from i = 1 to i = n − 1. In Burn's example cited above log LD_{50} is estimated as −1.773, resulting in an LD_{50} estimated at 0.59 cm^3/110 g.

Computing Confidence Bounds

It has long been realized that, owing to the randomness of the experimental data, estimates of parameters based upon them can be in error. Good methods of estimation, as defined by Fisher, make use of the experimental data to compute estimates that can be trusted as the best available. However, even the "best" estimates can be in error. It is thus a widespread practice to express the estimate of a parameters as a range of values. A

toxicologist might report, for instance, that the LD50 is 0.59 ±
0.06 cm^3/100 g. What does this mean and how are the calculations made?

The central limit theorem suggests that for a large number
of estimation procedures the random error associated with the
resulting estimate will tend to follow a Gaussian distribution.
So we can think of an estimate as consisting of two parts:

Estimate = true value + random error

where the random error has an underlying mean of 0, so we
need be concerned about only the second parameter of the
Gaussian distribution, the standard deviation. If we knew the
standard deviation of the estimate, we could report the estimated
value of the parameter and the standard deviation, and any
reader could then calculate a confidence interval whose width
would depend upon the degree of probability he wanted to take
into account.

The term *standard deviation* is somewhat ambiguous. We can
think of the underlying distribution of the data we observe as
having a standard deviation. If we used a good method of estimation, then the standard deviation of the estimate should be
less than the standard deviation of the underlying distribution.
That is, by running the experiment and combining the data in
an "optimal" fashion, we would hope to reduce the level of uncertainty. It has become common practice to refer to the standard deviation of the random error associated with the estimate
as the *standard error* or the *standard error of the estimate* to
distinguish it from the standard deviation of the original data.

Of course, we do not really know the standard error of the
estimate. If we did, we usually would have known enough about
the underlying distribution so that we would not have had to
run the experiment to begin with. So we estimate the standard
error of the estimate of the parameter. To the skeptical student
this will begin to sound like an infinite regression, requiring
that we find the standard deviation of the estimated standard
error of the estimate and then the standard deviation of the
estimated standard error, and so on.

However, a discovery of W. S. Gossett in 1908 gets us out
of this bind. When the estimate has a Gaussian distribution and
its standard error is estimated by the method of least squares,
then the ratio of the estimate to its estimated standard error has
a probability distribution that is independent of the parameters

of the underlying distribution. That is, we can stop with the estimate and its estimated standard error, since we can compute the probability associated with a confidence interval based upon the estimated standard error. Gossett published under the pseudonym of student, and the probability distribution of the ratio of the estimate to its estimated standard error is called *Student's t*.

The distribution of Student's t depends upon the number of "degrees of freedom" associated with the method of estimation. This is usually the number of observations less one degree of freedom for each underlying parameter estimated from the data. The conditions assumed by Gossett in deriving this distribution are fairly stringent: The estimate must have a Gaussian random error and the method used to estimate its standard error must be statistically independent of the method used to derive the original estimate. Soon after Gossett's work most applied statisticians came to believe that these stringent conditions were not really needed. In situations where the distribution of the random error was clearly not Gaussian or where the methods of estimation were not completely independent, the resulting confidence intervals often appeared to hold over repeated experimentation. This property of Student's t was called *robustness* by G. E. P. Box in the late 1950s. In the early 1970s B. Efron (1969) was able to find less stringent conditions (which he called orthant symmetry) for which the probability statements based on Student's t still hold true.

This approach, which uses the distribution of Student's t and the estimated standard error to construct confidence bounds on the true value of the parameter, has one very general extension that has become widely used now that high-speed computers are available: the *jackknife* (Miller, 1974). John Tukey named this procedure the jackknife because, like the young boy's jackknife, it can be used for all kinds of situations where a more specialized tool is not readily available. In the jackknife procedure we first calculate the estimate of the LD_{50} by some particular means (such as a Spearman—Karber method). Let us call the resulting estimate L_0. We then leave one animal out of the data and recalculate the LD_{50} by the same method; call this estimate L_1. We put the animal back into the data and leave out another, computing a second restricted estimate, which we call L_2. We continue in this fashion, constructing restricted estimates that result from leaving out one animal at a time from the calculations. If there are a total of n animals across all doses,

we then have the original estimate and n restricted estimates.
For the ith animal we compute the ith "pseudovariate":

$$M_i = nL_0 - (n - 1)L_i$$

This is the contribution of the ith animal to the original estimate
of the LD50. To see why this is so, think of the original estimate as the average of the contributions of all n animals. The
restricted estimate that results from leaving out the ith animal
can be thought of as the average of the contributions of all the
other animals. We multiply the average of all the animals' contributions by n and get the sum of all their contributions; we
then multiply the average of all but the ith animal by n − 1 and
get the sum of all their contributions. Subtracting the second
quantity from the first leaves only the contribution of the ith
animal. This description of the pseudovariates holds exactly if
the original estimate is the average of all the observations.
When the original method of estimation is more complicated (as
in estimating the LD50), it does not hold exactly; however, it
has been shown in the theoretical literature that the pseudovariates get closer and closer to being independent as the number
of observations increases, as long as the method of estimation
follows certain regularity requirements (Arvesen, 1969).

If, in addition to having a proper method of estimation, the
pattern of pseudovariates has Effron's orthant symmetry, we
can then use Student's t distribution to construct confidence intervals on the parameter being estimated in the following fashion.

The average of all the pseudovariates is taken as the final
estimate of the parameter. (In fact, if the original method is
biased, the average of the pseudovariates is less biased.) We
then compute the differences between the average of all the
pseudovariates and the individual variates. The average of the
squares of these deviations is an estimate of the standard error
of the jackknifed estimate, and the jackknifed estimate and its
estimated standard error can be used with Student's t distribution to compute confidence intervals on the parameter.

The jackknife will always produce a less biased estimate than
the original procedure. Jackknifed bounds, on the other hand,
could be in error if the pseudovariates do not fulfill Efron's
condition of orthant symmetry. One condition that leads to orthant symmetry occurs as follows: We take the deviations of
the pseudovariates from their overall average. Some of these
deviations will be negative (the pseudovariate is less than the

average) and others will be positive (the pseudovariate is
greater than the average). If the occurrence of a + or − sign
is independent of the size of the deviations, then the set of
values has orthant symmetry. So a quick check on this condition can be made by looking at the pseudovariates: If many of
the very small or very large deviations all have the same sign,
then the condition may be violated.

It may be, on the other hand, that a small mathematical
trick will bring it back. For instance, when the LD50 is estimated, it is usually done first on the log dose metameter. Thus
we estimate the logarithm of the LD50 and then convert that to
the estimated LD50. It may turn out that jackknifing log LD50
rather than the LD50 itself leads to a more symmetrical distribution for the pseudovariates, or the LD50 may be a better
choice than log LD50. A toxicologist who uses the jackknife
to compute confidence bounds would do well to examine the
scatter of pseudovariates for both cases.

More on Confidence Bounds

The methods of computing confidence bounds on the LD50 described in the previous section are all based upon use of the
central limit theorem. When the LD50 and the slope of the linear portion of the sigmoid are estimated by any of the least-
squares method (linearized or not), the usual computer algorithms will also compute standard errors for those estimates based
upon the standard deviation of the residual values of the form

    Observed data − predicted P(death)

The resulting calculations of bounds on the underlying parameters require that we assume that the underlying distribution
function belongs to the family (probit, logit, or angle) that we
have chosen. Often the width of the confidence interval will
be more dependent upon these assumptions than upon the data.
The toxicologist should examine a plot of the observed data on
the same graph with the best-fitting sigmoid from the family
chosen and a plot of confidence bounds on that curve. If most
of the data do not fall within these bounds or if they do not
follow the path proposed by the best-fitting sigmoid, the toxicologist should be suspicious of the calculated bounds.

Finally, if the method of maximum likelihood is used to estimate the LD50, it is theoretically possible to compute the exact

standard error of the maximum likelihood estimate by applying calculus to the mathematical form of the underlying probability distribution. However, the toxicologist who uses computer programs that calculate estimates of the LD_{50} by maximum likelihood should recognize that there are at least two different methods for calculating the standard error of the estimates and that they produce slightly different answers.

Testing Goodness of Fit

When one particular mathematical model or family of distributions is chosen as a basis for estimating the LD_{50}, it is often useful to have a check on how well the data actually fit the best-fitting sigmoid from that family. Goodness-of-fit tests are a measure of the degree to which the observed data differ from the expected values based on the best-fitting sigmoid. Such a measure is itself a random variable, since it is based on the random error in the original data. As a random variable, it has a probability distribution. The measure of disagreement is called a *test statistic*, and most standard test statistics were created so that the probability distribution of the test statistic is known, almost regardless of the underlying probability distributions that may be true.

The Litchfield—Wilcoxon article (1949) proposes the use of a chi-squared goodness-of-fit test statistic that was developed by Kearl Pearson in the late nineteenth century. If the model claims that

$$P(death) = f(dose;\ slope,\ LD_{50})$$

and if the slope and LD_{50} have been estimated from the data, then the predicted probability of death can be compared to the observed proportion of animals dead at each dose,

$$(Number\ of\ animals\ dead)/(number\ of\ animals\ exposed) = x/n$$

The difference between the observed and the predicted proportions is divided by an estimate of the standard deviation of that difference, based upon the best-fitting sigmoid. These standardized differences are then squared and summed. The sum of these is a chi-squared test statistic whose probability distribution depends only on the number of doses used and the number of parameters estimated.

Litchfield and Wilcoxon proposed using the chi-squared goodness-of-fit test statistic as a check on the validity of the probit model. If the test statistic is "significantly" large, it means that the observed data deviate from the predicted by a degree that is highly improbable if the data really came from a Gaussian distribution. The general thinking behind such significance tests is that we do not tend to observe improbable events. Since we observed this event, its improbability under the Gaussian assumption suggests that the Gaussian assumption may be in error. Thus the goodness-of-fit test provides a warning that we may be imposing an inappropriate model on the data and that our resulting estimates of the LD_{50} and its bounds may be wrong. From a practical point of view, a failure of the data to fit the Gaussian distribution may not introduce much error in the estimated LD_{50}, since values very close to it would be estimated when other models are used; however, the confidence bounds are based heavily on the Gaussian assumption, and so the toxicologist might seek to compute confidence bounds based on less model-bound procedures, like the jackknife or the Spearman−Karber method.

The chi-squared goodness-of-fit test proposed by Litchfield and Wilcoxon is not the only way to test goodness of fit between the observed and the predicted. The chi-squared test is a general one. It examines the degree of deviation of the observed from predicted in an absolute fashion on the assumption that any other distribution might fit. By allowing for all possible distributions, it tends to lack power; that is, it is not capable of detecting slight deviations that suggest other models which are not far removed from the Gaussian. In statistical theory it is possible to increase the power of a test statistic by limiting the alternative possibilities. The more limited the alternatives, the more effectively the test statistic is able to determine when the proposed model is inadequate. We can construct a test statistic that is not so general as the chi squared if we can find a class of distributions to test against that that is slightly larger than the Gaussian.

The most widely used test statistic of this kind is the test for linearity. Recall that the linearized least-squares method of estimating the slope and the LD_{50} was based upon the equation

$$M(P(death)) = A + B(dose)$$

A little more flexibility can be introduced by proposing the equation

$$M(P(death)) = A + B(dose) + C(dose)^2$$

This has three parameters: A, which is determined by the
LD50, and B and C, which allow for some flexibility in the sigmoid curve. We compute the best-fitting values of A, B, and C
under the second model and the best-fitting values of A and B
under the first model. (Since the second model has a third parameter to account for some of the random noise, the estimates
of A and B need not be the same for the two models.) We then
compute the deviations between observed and predicted for the
first model and again for the second model. The sums of the
square deviations are compared in a test statistic invented by
Fisher in the 1920s. If the observed value of this test statistic
(usually called an F test) is significantly improbable, then we
tend to reject the two-parameter model in favor of a three-pa-
rameter one. The function M(P(death)) is still based on the
Gaussian distribution in both models, so this does not test
against the possibility that the underlying distribution is not
Gaussian (logit or angle), but it will tend to reject the assumption of a two-parameter probit for sets of data that the chi-
squared goodness-of-fit test would not.

Unit II
Subchronic Toxicity Studies

4

Subchronic Toxicity Studies

Overview

The Food and Drug Administration of the U. S. Department of Health and Human Services identifies two types of subchronic studies in their general guidelines for toxicology testing: the short-term continuous-exposure oral toxicity study, which lasts for 1 month or less, and the subchronic oral toxicity study, which lasts from 90 days to 12 months. The short-term study has a more limited objective (the identification of target organs of toxicity), but the general design of the two types of study is similar and the statistical techniques used are identical.

In these studies groups of test animals are continuously fed various doses of test substance. The doses are chosen so that there is one control group, one group on a high enough dose to produce manifest toxicity, and two or more intermediate dosing groups, at least one of which is designed to be so low as to have no apparent adverse effect. The animals are assigned at random to treatment groups and the positions of the animal cages in the holding room are randomly assigned with respect to treatment. Treatment begins while the animals are still juvenile (soon after weaning for rodents and 3–5 months old for dogs). Observations are made on animals both while alive and after death.

In order to understand where and how statistical methods are used in the analysis of these studies, it is useful to outline those components that call for some sort of statistical treatment:

1. Assignment of animals at random to treatment
2. Observations made on living animals
   a. Patterns of growth chemistries
   b. Clinical signs and chemistries
3. Observations made upon death
   a. "Cause of death"
   b. Gross pathology
   c. Histopathology

In the guidelines for both types of studies, the Food and Drug Administration (FDA) notes that

> an evaluation of test results, including their statistical analysis, should be made and supplied, based on the clinical findings, the gross necropsy findings, and the histopathological results. This should include an evaluation of the relationship, or lack thereof, between the animal's exposure to the test substance and the incidence and severity of all abnormalities; such abnormalities include behavioral and clinical abnormalities, tumors and other lesions, organ weight effects, effects on mortality, and any other general or specific toxic effects.

This paragraph defines the specific components of the analysis of data that should be considered. The overall purpose of these studies can be thought of in terms of their use in the next phase of the development of a new chemical compound, which is to specify the following:

1. A starting dose for the next phase
   a. A "safe" dose for first human exposure if the compound is a drug or food additive
   b. A "maximum tolerated dose" for planning the design of a chronic toxicity study

2. The nature and characteristics of induced organ toxicity and a dose response in terms of incidence and severity.

In what follows we will examine how well statistical techniques are able to make these two overall determinations (the starting dose and the nature of organ toxicity) and to what extent it is possible to evaluate "the relationship. . .between. . .exposure. . .and incidence and severity" of toxic manifestations.

Random Assignment of Animals to Treatment

Since the early 1920s it has been recognized that comparisons of two or more treatments require that the experimental material be assigned at random to the treatments. This is true whether this concerns an agricultural experiment (and the treatment material consists of small areas in a field), human clinical trials, or toxicity trials.

There are two ways of viewing this experimental requirement. From one point of view, we think of the response of the animals as reflecting more than just the experimental treatment. The response might involve subtle differences in genetics, the contents of the animals' stomachs, the lighting or air patterns in particular parts of the holding room, and so on. Since we cannot characterize or even measure all of these possible variables, we assign the animals to treatment at random so that, over a large number of animals, these extraneous effects average out. That is, if we assign a large number of animals at random, those whose uncontrolled factors will cause an increase in the incidence of some naturally occurring lesion will be balanced out by others who have a built-in resistance to that lesion in any of the treatment groups.

This first point of view, that randomization is needed to "balance" the groups, leads the toxicologist to examine how well the groups are balanced at the beginning of the trial. Thus some toxicologists will check on whether all groups have the same or nearly the same average weight and that the baseline measurements of blood chemistries "balance out." If they do not, some people will rerandomize. In order to ensure a better balance some toxicologists will prune the animals to only those whose weights and blood chemistries are within a narrow "normal range" before randomly assigning treatments.

In laboratories where rodents entered into a trial are bred on the spot from pregnant females, it is the usual practice to

allow the litters to live for a week or so after weaning. Only
the most robust pups are then used, and the weaklings are
culled out. Sometimes an attempt is made to balance the study
by keeping enough animals in a litter so that at least one can
be assigned to each treatment. For those who believe that
balancing is the main purpose of randomization, there is often
the temptation to "balance" the assignment better than would
have been done at random by reassigning animals on the basis
of some complicated examination of their baseline conditions.

At this point the proponents of randomization as a means of
balancing run up against the other view of randomization. This
second view is more subtle, and it is based upon the mathematical theorems behind the methods of statistical analysis usually
applied to the data from such studies. In order to apply the
standard methods of statistical analysis, the normal pattern of
variability that might occur among untreated animals must be
estimated. The degree to which the average patterns of "response" differ between two treatment groups is then compared
to the estimated normal pattern of variability. If the average
patterns of two treatments differ by more than would be predicted by the normal pattern of variability, then the treatments
are declared to have had different effects.

At the heart of this method of analysis is that estimate of the
normal pattern of variability. The act of randomly assigning
animals to treatment ensures that the standard procedures will
correctly estimate the normal pattern of variability. If the animals have been "balanced" by some other means, the standard
procedures will tend to underestimate the normal pattern of
variability and things which may not differ by more than chance
will then appear to be significantly different.

A philosophical problem emerges from the contrast of the two
views: What happens if a "bad" randomization occurs? That is,
what should be done if after random assignment of animals to
treatment it is obvious that they are not "balanced" with respect
to some measure (such as weight)? The first view says rerandomize; the second says take the randomization as it fell. The
Talmudic discussions that have arisen from this contrast in views
can be found in both statistical and toxicological journals.

There is, however, a statistical answer to the problem. If
one discovers an "imbalance" in one or more of the baseline measurements of the treatment groups, the statistical analysis can
adjust for that imbalance by a method known as *analysis of co-*
variance, which is also based on the assumption that assignment
to treatment was random. The usefulness of this method of analysis and its pitfalls will be discussed later in this chapter.

Although it is assumed throughout this book that the student has access to computer programs to run standard statistical analyses, it would be useful here to examine one method of random assignment. For many who have not studied statistics randomization seems to be a mysterious process and is sometimes confused with complex methods of assignment.

We assume that the animals arrive in a "box" in such a fashion that they can be taken out of their container one at a time. They have to be assigned to cages in a holding room so that they can be assigned at random to treatment and the treatments assigned at random to cages in the holding room. In some sense, this is a two-step randomization procedure, however, theoretically, there should be no difference in whether one shuffles a deck once or twice. So a simple procedure is to place the animals in the holding room as they are taken from the box in any fashion whatsoever. Next we consult a table of random numbers and list the cages in the holding room in order against the sequence of random numbers as follows:

Cage number	Random number
1	19
2	39
3	64
4	92
5	03
6	98
7	27
8	59
9	91
10	83
⋮	⋮

The random numbers are then put in ascending order, paired with their cage numbers. If in the above example we have stopped at cage 10, we would obtain

Cage number	Random number
5	03
1	19
7	27
2	39

Cage number	Random number
8	59
3	64
10	83
9	91
4	92
6	98

The 10 cages have now been given a random order, 5,1,7,2,... .
If we have three treatments—A, B, and C—we list them in
order against the randomly ordered cages:

Cage order	Treatment
5	A
1	B
7	C
2	A
8	B
3	C
10	A
9	B
4	C
6	A

So treatment A is now assigned to cages 2, 5, 6, and 10; treatment B is assigned to cages 1, 8, and 9; and treatment C is assigned to cages 3, 4, and 7. Note that there has been a single
shuffle of the deck; the random number table has been consulted
only once.

In spite of the earlier injunction against attempting to "balance" the treatment assignments without randomizing, there are
methods of balancing the randomizations to take into account obvious differences. The most widely used method is to consider
the holding room as subdivided into racks, each rack holding
a fixed number of cages. For rodent studies these racks of
cages often consist of five or six rows and four or five columns,
with two cages back to back in each position. The racks are
often on wheels to allow for the easy transfer of animals to a
weighing station and for convenience in cleaning.

If the racks are thought of as the units of experimentation,
it seems reasonable to balance a study so that all treatments are

randomly assigned to rack position, with each treatment equally represented in a given rack. This is done by limiting the number of animals in a rack to some multiple of the number of treatments. Suppose there are five rows and four columns of cages back to back, that is, 40 cages to a rack. Suppose the study has four treatment groups and two controls, one a vehicle control and the other a nontreated control. This means a total of six treatments to be randomly assigned among 40 cages. If we assign six cages to each treatment, we have 36 (6 × 6) cages that will be used, with four left empty. As we did above, we list the 36 cages in order, rank them alongside 36 numbers taken from a random number table, reorder the 36 cage numbers, and assign the treatments in order to the randomly reordered cage numbers. If we had a rack with six cages per column, we might have wished to subdivide the assignment such that all treatments were represented in each column. Then we would have listed the cages in each column against a list of random numbers and reordered them. We would have had to use a different list of random numbers for each column so that the same treatment did not always occur in the same row.

At the other extreme of "balanced" but randomized schemes is one in which an entire rack is given over to a single treatment and the racks are assigned at random to treatment. Assigning an entire rack to a single treatment is convenient for handling. There is little chance that an animal will be given the wrong treatment if all the animals in the rack get the same treatment. However, with modern computer methods this is no longer such a convenience, for most toxicology computer programs will print out labels for each cage indicating the current week's feeding for that treatment. In some institutions these labels are put onto individually prepared feeding bags for each cage and the chance of misassignment is thus greatly reduced.

The toxicologist should keep in mind that all of these balanced randomized schemes carry a further burden in analysis. The formal statistical analysis of the study should include certain factors to account for the balancing. It will usually turn out that these factors are negligible in their effect on the final conclusions when the treatments are balanced within racks or columns, but the standard method of estimating normal variability has a provision for such detailed balancing.

The toxicologist should also recognize that gross forms of balancing such as the one where each rack is assigned to a different treatment lead to a problem called *confounding*. Two

aspects of an experiment are said to be confounded when the
effects due to one aspect are impossible to determine without
combining them with the effects due to the other. If each rack
has a different treatment, the treatments and racks are confounded; any apparent difference in effect between two treatments could be due to true differences between treatments or to
differences between the environments of the different racks.

Observations on the Growth of Living Animals

Since the animals in a subchronic toxicity study begin treatment
while still immature, an important part of the study concerns
the effect of treatment on their growth. Although it might be
possible to measure "growth" in terms of the animal's length,
the maturity of its sexual organs, or some other esoteric concept, growth is usually measured in terms of weight. For
larger animals, such as dogs, monkeys, and rats, the individual animal is usually measured at weekly or biweekly intervals.
For smaller animals such as mice the entire cage containing
three to five animals is often weighed and a total or average
weight computed by subtracting the weight of the empty cage.

It used to be the practice in some labs to weigh randomly
selected or typical animals from each treatment group. This
led to difficult statistical problems when the same animals were
not weighed each time, and the advent of automatic scales directly linked to the computer has eliminated any convenience
that less than complete weighings might have had. So in this
discussion we will assume that each individual animal still living
is weighed at each point in time.

Each study thus produces a large collection of weights. In
a typical 90-day study, with weekly weighings, three treatments,
and a control group of 20 rats per sex, almost 2000 pieces of
data are collected for weights alone. It is tempting to analyze
these numbers merely because they are so numerous and very
orderly. However, any statistical analysis should be directed at
a well-defined toxicological question, and some of the more interesting "statistical" analyses of such data are really peripheral
to the purpose for which they were collected.

There are three general toxicological questions that one
should attempt to answer with these animals weights. The first
deals with the running of the trial. If the animals are still
growing and if the protocol defines the treatment in terms of
the dose per kilogram of weight, the toxicologist needs to pre-

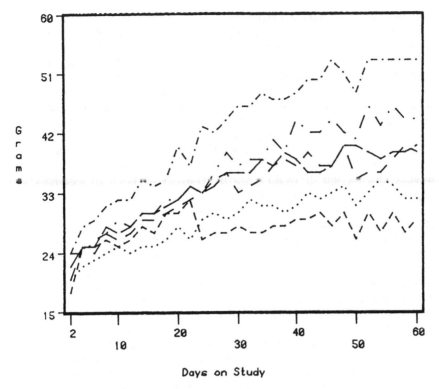

Figure 1 Weights of individual mice.

dict the average weight of each treatment group of animals over the next week in order to mix the feed with the proper proportions of test chemical.

Individual animals tend to grow in unpredictable patterns. Figure 1 displays growth curves (weights at each week) for individual mice from a 90-day toxicity study. In some cases an animal might go off its feed for 1 week and the influence of that event will be apparent for many weeks as that animal maintains a weight below the average, or a sudden spurt of weight may cast its influence. In general, however, there is an upward trend influenced by random individual events. A great deal has been written in the statistical literature on growth curves and the toxicologist literature has tried to characterize the average growth patterns of each species in terms of mathematical functions. Attempts have been made to use these insights in predicting weights for the next week.

However, there is a simple procedure that appears to work most of the time. It is based upon the general principle that what will happen in the next short interval of time is not much different from what happened in the last short interval of time. Thus the average weight gain for a group of animals during the next week can be expected to be very similar to the average weight gain observed over the past week. This can be calculated by either adding the average gain of the past week to the current weight or computing the ratio of this week's average weight to that of the previous week and multiplying this ratio by the current week's average. The ratio approach works best during the period of rapid growth; thereafter both procedures appear to work equally as well.

The current practice is to calculate the amount of treatment chemical to be mixed in the feed for an entire treatment group, based on predicted average weights. However, we now have computers that keep track of the weights of individual animals (or cages, in the case of mice). So, it is theoretically possible to calculate the amount of treatment chemical for each animal (or cage) in a group. In this way, an animal now weighing 65 grams will get less chemical than one weighing 75 grams, and all animals in a group will get the same mg/kg/day dose. However, this may be an example of fine tuning that is an unnecessary refinement. Every experiment carries some residual degree of uncertainty, which is the sum of the random noise due to a large number of uncontrollable factors. The statistical analyses of toxicological experiments suggests that the residual degree of uncertainty due to other factors is so great that more exact knowledge of the "true" dose of treatment will not improve matters much.

The second toxicological question raised in keeping track of animal growth deals with the choice of doses to be used in chronic, long-term dosing studies. Current guidelines call for use of the "maximum tolerated dose." This is usually defined as the dose that produces a 10% reduction in weight.

Actually, the concept of a 10% reduction in weight is much too vague to be applied directly. What does it mean? Should the average weight in the treated animals at the end of the 90-day study be 10% below that of the controls? If so, is this 10% of the average control weight, the average of all weights (treated and controls), on the average of the treated weights? Should the choice be based upon a formal statistical test of the hypothesis that the difference in mean weights is less than 10%? And, again 10% of what? Does it mean that a theoretical curve

of weight versus time should be fit to the data for the treated
and control animals and the endpoints of these curves compared?
Each formulation of the vague question will lead to a slightly
different answer, and the treatment that induces a "10% reduction in weight" by one definition may not do so by another.

We will not pursue this question any further, since the toxicological community has not come to a specific consensus as to
the meaning of this concept. Actually, the choice of doses for
a chronic feeding study is usually based upon an overall examination of the clinical toxicity observed (weight, blood chemistry,
behavior, etc.), and the choice of a "maximum tolerated dose"
is often not any of the doses used in the 90-day study but,
rather, one interpolated from the dose data obtained. Furthermore, if it becomes obvious, as the chronic study gets under-
way, that the dose chosen as "maximum tolerated" was too high,
the dose is usually lowered. If there is any doubt about whether the dose used in a chronic study was the maximum tolerated
dose, the issue is usually addressed by looking not at the prior
90-day subchronic study but at the data that emerge in the
long-term chronic study itself.

The student should keep in mind that there are other equally
vague concepts in toxicology. These concepts are left vague
and cannot be addressed by a structured statistical approach.
However, judgments can be made that seem to answer the
questions posed and upon which most toxicologists will tend
to agree. As long as a consensus can be reached, it would
seem inappropriate to concoct a formal statistical procedure and
force the vague concept into a more rigid (and possibly incom-
plete) formalization.

The third set of questions answered by keeping weights is
referred to in the FDA guidelines quoted earlier. The weight
is part of the overall picture of an animal and can be used in
evaluating the relationship between exposure to test compound
and any abnormalities observed.

When a statistical method is used to help answer questions,
it is necessary that the question be very clearly defined. Once
a well-defined question is asked that has a specific numerical
answer, there will usually exist a standard statistical procedure
that will be adequate to answer the question. The hardest part
usually is finding a well-defined question. In what follows we
examine some possible questions about relative animal growth.
Methods of statistical analysis are postponed to a later section.

As the subchronic study progresses each animal is weighed
at regular intervals until death. For a particular animal, say,

the ith animal, we can represent these weights as a string of numbers

$$x_{i0}, x_{i1}, x_{i2}, x_{i3}, x_{i4}, \ldots, x_{it}$$

where t is an index of time. If the animal died during the study or was killed early when moribund, t will be a number less than the number of weeks planned in the study; if the animal lived to terminal sacrifice, t will be the number of weeks planned in the study.

One way to compare two treatments is to consider the weight gain from the start of the study to a fixed point in time. We could, for instance, ask whether there was a treatment effect by week 4. Then we might obtain the average difference between the baseline weight and the weight at week 4,

$$x_{i4} - x_{i0}$$

for all the animals on the first treatment and compare that to the average change from baseline to week 4 for all the animals on the second treatment. Comparing the average weight gain tells us something about the general pattern of "response" across all animals. However, there are times when the toxicologist is interested in extreme, deleterious events. Suppose that the compound being tested is suspected of inducing serious early toxicity at high doses that will cause a decrease in weight gain or even a loss of weight but only in the weaker animals. Then it might make more sense to count the number of animals that fail to gain more than 20% of their baseline weight by 4 weeks and compare these counts between treatment groups.

If it seems best to compare overall, average patterns, the use of average changes from baseline after a fixed interval of time leads to a practical problem. How do we use the data of animals that died before that interval of time ended? If the comparison is between baseline values and those at the end of the study, we have an acute problem, since 10--20% of the animals may have died. Usually, these are animals that have been administered the highest dose, the one planned to have a toxic effect. If we ignore these animals, the survivors at that highest dose will tend to have less of an effect on the data than those that succumbed to the toxicity. If, however, we decide to take into account the animals that died, what weights should be used? We cannot use the weights at the time of death, since this might include juvenile weights and would lower the apparent toxic doses.

One way around this problem is to find some overall measures of weight change that can be computed independently of whether the animals lived to the final sacrifice or not. If we think of the growth curve as having a rapidly rising juvenile phase followed by a slow, steady growth phase, we can fit straight lines to the latter period for each animal. The slope of that line for a given animal is a measure of the change in weight over the period of the study, and it can be computed for every animal that lives to the point of slow, steady growth, regardless of whether it lived to terminal sacrifice. Thus these slopes can be used to compare the treatments using all the animals assigned to treatment.

Similar measures for comparison, using all the animals, are the average weight while on treatment, or the area under the time curve, which is a measure of cumulative growth.

The previous paragraphs provide some examples of the choices the toxicologist can make among the measures of effect when comparing the growth (in terms of change in weight) of animals subjected to different treatments. Since there is such a large number of possibilities, the toxicologist might want to first examine the average weight across time and the percentage of animals that die or are killed before terminal sacrifice. Thus a good starting point in the examination of this large amount of data is a computer printout that lists the average weights at each observation time and the counts of the numbers of animals remaining. It is common practice to have the computer also run simple statistical tests of whether the differences in average weight among the treatments are sufficient to suggest "significance." Table 1 displays such a printout, with "significant" differences noted.

While listings of average weights and numbers of animals remaining on trial are useful as a first examination of whether and to what extent each treatment induced a toxic effect, the "significance" pattern may lead the toxicologist astray. Statistical tests that can be run on such listings of data tend to produce false positive and false negative results in unpredictable ways. Since such printouts of patterns of "significance" cannot be interpreted in the same way as formal statistical tests of significance usually are, they should be used as no more than a guide to the toxicologist. For instance, a pattern of "significant" depressions in weight during the beginning of the study only,

Table 1 Average Weights (g) of Female Rats

		Dose level			
Week	Controls	5	4	3	2
5	236.7	221.6	221.2	232.1	237.3
6	243.6	228.7	227.8	242.1	247.2
7	250.5	237.2	239.2	252.9	275.0
8	260.6	245.0	245.1	256.5	269.1
9	271.6	259.0	253.2	259.5[a]	170.5
10	275.4	256.4	256.6	271.7	274.6
11	281.8	257.0[a]	253.1[b]	272.2[b]	281.5
12	289.0	261.9[b]	260.2[b]	264.7	287.8
13	286.8	262.4[b]	261.0[b]	276.3	285.7
14	288.5	262.2	262.5[a]	278.9	287.7
15	293.4	267.4	266.1	280.6	279.7
17	306.1	266.5[b]	273.1	285.8[b]	298.4[b]
18	302.2	265.1[b]	274.5	288.8	301.3
19	315.1	265.3[b]	283.0	293.8[a]	306.1
20	310.2	263.3[b]	283.4	293.7	306.3

Key: [a]$p < .05$

[b]$p < .01$

with only slight changes in average weight, might suggest that
almost all treated animals fell behind the controls, possibly the
result of the treatment adding a bad taste to the food. At
other times purely random noise can introduce apparent patterns.

More Observations on Living Animals

At the beginning of a trial and at the end, coincident with the
final sacrifice, the animals are given an ophthalmological examination and blood is subjected to hematology and clinical
chemistries. In addition, daily observations are made for signs,
such as biting reflexes. In most labs moribund animals are
culled so that the final measures can be made before they die
and undergo autolysis or are cannibalized. Some of these measures lend themselves neatly to "statistical analysis" and more
or less standard methods of analysis have been developed.

Table 2 SGOT (mU/ml) Values for Male Rats—Week 0 to Week 26

Week	Controls		Dose 1		Dose 2		Dose 3	
	0	26	0	26	0	26	0	26
	176	272	142	154	164	265	199	195
	185	175	140	123	157	232	164	272
	182	155	166	108	150	206	155	259
	183	155	153	179	169	224	157	320
	158	155	163	121	174	186	182	255
	162	162	162	155	142	187	159	311
	150	179	142	169	129	258	150	265
	204	179	132	169	170	237	156	254
	160	165	142	124	132	251	163	286
	162	196	181	236	138	269	225	323
	157	162	186	168	123	175	140	266
	165	156	162	168	158	267	157	278
	211	161	143	150	162	238	169	280
	150	190	146	145	152	202	200	304
	136	141	167	125	163	214	149	246
	165	165	162	83	143	222	130	265
	158	173	170	119	158	278	199	234
	176	171	167	166	172	178	124	268
	123	169	152	139	155	176	191	259
	136	149	152	105	164	251	171	254
Average	165	166	156	142	154	225	167	270

There may be some doubt concerning the usefulness of these analyses, but they are easily run on the available data and therefore tend to make up a large part of the reports generated from such studies.

To illustrate these methods, let us concentrate on a single variable, the serum glutamic-oxaloacetic transaminase (SGOT) level as a measure of liver damage. For each animal at least two SGOT values are obtained: a baeline value, taken at the beginning of the trial, and one value taken at the end of the trial. For those animals that are killed or die early the second SGOT value will be taken closer in time to the first than for those animals that lived till the planned final sacrifice.

We can list the pairs of values—SGOT level and time—for each animal, as in Table 2. Statistical analysis of this type

of data starts with these individual animal values. The goal of
the analysis is to reduce long lists of data to some form that
can be interpreted at a glance and easily understood and which
carries the essential information needed to reach a conclusion
on any differences in effect among the treatment groups. One
obvious thing to do is consider the arithmetic differences between the baseline and the final values. The SGOT levels can
be expected to have changed, since all the animals (both controls and treated) will have gone through a period of change
and growth. If, however, all the changes for a particular
treated group are greater than all those for the controls, one
would tend to believe that the treatment had some effect that
was reflected in the SGOT values. If all the changes in a
treated group overlap in value among those seen in the control
group, we would tend to dismiss the possibility that the treatment had an effect reflected in the SGOT values.

This is the rough idea behind the formal statistical analysis
of such data. Does the scatter of differences for a treated
group belong to a different distribution than the scatter of differences for the control group? This question can often be answered by simply plotting the values against each other. Figure 2 displays the changes in SGOT for two groups of 20 male
rats each exposed to different doses of test compound and a
group of 20 controls. To aid the eye in examining Figure 2, the

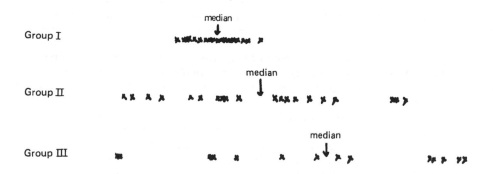

Changes in SGOT Week 0 to Week 26

Figure 2 Changes in SGOT, variance increasing with mean.

median change (the one in the middle of the range of observed values) has been indicated for each group. A gradual increase in the median change can be seen across doses and the last dose group has a scatter pattern that is clearly to the right of the scatter pattern of the control group.

This example can be formalized in a statistical procedure known as the *Kruskal–Wallis nonparametric ANOVA test* (Hollander and Wolfe, 1973). As with the goodness-of-fit tests discussed in Chapter 1, we assume that the toxicologist has a computer program that produces two numbers:

1. A test statistic (usually called chi squared)
2. A significance level (p value).

The two are equivalent, so we can concentrate our attention on the p value. The p value is the probability of observing such scatter (or any scatter deviating more than this from the equality of patterns) when there is no difference between the underlying probability distributions that produced the four sets of data. If the calculated probability is low, we are faced with two alternatives:

1. We have observed an event that is improbable.
2. There is something wrong with our calculations.

Since we do not usually observe improbable events, we are forced to choose (2) as the more likely explanation. What could be wrong with our calculations? Surely there is nothing wrong with the mathematics of Kruskal and Wallis! In fact, there is only one part of the calculations that could be wrong, and that is the initial assumption that the treatments had no effect on the SGOT levels.

This, then, is the essence of *statistical hypothesis testing*. We propose a hypothesis which we expect to prove wrong ("no treatment-related effect" is the usual one). We calculate the probability of what we observed plus the probabilities of any possible observations that would be even less probable under the hypothesis of no effect. If this results in a very small probability, we conclude that we could not have observed something improbable, so the initial hypothesis must be wrong.

Note that a statistical test of hypothesis cannot be used to show that the initial hypothesis is correct. Just because we have observed something that is not improbable under a given hypothesis does not mean we can conclude that the hypothesis is true. There are an infinite number of other hypotheses that could also be true that would produce reasonably high probabilities for our observations.

The Kruskal–Wallis test has been used to illustrate the concept of a hypothesis test as applied to differences in SGOT between baseline and final observations across treatments. The calculation of Kruskal–Wallis probabilities is a complicated and tedious procedure. During the 1930s, before the advent of modern computers, these calculations would have been too much for a desk calculator, so methods were developed that enabled statisticians to compute probabilities by other means. The most widely used procedures were based upon assuming that the data fit a Gaussian distribution (remember Bliss's probit curve from Chapter 1, which reflected this assumption). This assumption was chosen not so much because it was thought to be true, but because it led to mathematics that could be done easily on a calculator. The basic tool is called *analysis of variance*, and abbreviated *ANOVA*.

(In fact, the Kruskal–Wallis test was invented after the original ANOVA. Because it deals with the same type of problem, it is called a nonparametric ANOVA procedure—nonparametric because it does not assume a particular parametrized probability distribution.)

Standard ANOVA produces an answer of the same type as the Kruskal–Wallis test. It calculates a test statistic called an *F test* (in honor of R. A. Fisher, who invented the procedure) and an associated probability of observing a value of the F test greater than or equal to the one observed. If this probability is small, we dismiss the hypothesis that there was no treatment-related effect.

The use of ANOVA (Gaussian or nonparametric) was shown as applied to differences between baseline and final SGOT values. In fact, the same statistical tests can be applied to any measures of effect. In the previous section we noted several measures of growth that might be compared. The student should recognize that a standard statistical test such as ANOVA is really an empty glass into which one pours the "wine" of data. The glass is indifferent to the type or quality of wine it is asked to hold. It is up the scientist who uses it to taste the wine to see that it is appropriate for the "dinner" being prepared.

Statistical Tests of Hypothesis Reexamined

In the previous section it was noted that the only way in which the calculations of the probability could be wrong would be if the initial hypothesis of no effect were wrong. Strictly speaking, this is not true. We can be sure that the mathematics of Kruskal and Wallis or of Fisher are correct as far as they go, but there are other more subtle assumptions involved besides the initial choice of hypothesis. The Fisher ANOVA procedures assume that the data being examined came from the same Gaussian distribution. The Kruskal—Wallis procedure assumes that they came from either the same probability distribution (the null hypothesis we have already examined) or else a set of probability distributions that differ only in their medians but have the same type of scatter.

What all this means is that it is possible for the treatments to have effects that distort the probability distributions (changing them from Gaussian or modifying them in more ways than just in their medians) enough to make the differences in pattern obvious to the eye. However, the distortion may be of a sort that the mathematics of both procedures did not anticipate. As a result, differences that may be clear to the eye will not produce improbable test statistics with Fisher's or Kruskal and Wallis's ANOVA.

In fact, Figure 2 displays just such a pattern. Here as the dose of treatment increases, the scatter of values increases. Such an "effect" can emerge in a toxicity study if one is measuring some value (like the SGOT level) where variability increases as the median increases. There may be more subtle events that can occur and which will fool the usual statistical tests. One way around such problems is to use a *transformation* of the data rather than the original data recorded. The mathematical theorems behind the use of transformations enable the statistician to suggest modifications of data that should eliminate many problems. Thus, when the toxicologist observes differences that appear to be important but for which the p value is not small, he or she should consult a trained biostatistician, who might be able to recommend a useful transformation.

Up to this point we have carefully avoided two aspects of statistical hypothesis testing which are fundamental parts of the jargon and procedures as now used. It is time to introduce them, albeit with a grain of salt. Kearl Pearson was the first to formalize the concept of statistical hypothesis testing. In doing so, he made the unfortunate use of a word which had a limited meaning in his Victorian upper-class English. He proposed that a low computed probability be called *significant*, meaning nothing more than it "signified" that something has happened different from the proposed hypothesis. As a result, hypothesis testing is sometimes called *significance testing*, and results are declared to be significant and nonsignificant or statistically significant and non-statistically significant.

Unfortunately, the word *significant* has come to mean more than Pearson intended. In modern American English the word tends to mean something important or wonderful, so there is a tendency to interpret findings of "statistical significance" as being important by themselves. A significant probability is evidence against the assumption that there is no treatment-related effect. It need not mean that the effect (if there is one) involves toxicity. It does not reflect the degree of effect either, since a finding of significance is based on both the degree of effect and the underlying scatter of data. If there is very little scatter, a minor degree can be significant. For example, rodents treated with chronic doses of aspirin will experience a slight rise in blood urea nitrogen levels, even at the lowest doses used. Almost all the treated animals, but very few of the control animals, will have such increases. The result is a slight but consistent rise associated with low variability but statistical significance against the hypothesis of no treatment-related effect. What does this mean to the toxicologist who is seeking to profile the toxic nature of the treatment?

The second aspect of statistical testing (already referred to in Chapter 1) that has become part of the standard jargon is the choice of what levels of probability are "improbable." That is, how do you decide that the calculated probability is significant? Certain arbitrary choices have become widely used. The reader should keep in mind, however, that these choices are, in fact, purely arbitrary. It has become customary in most areas of science to call p values between 1 and 5% "significant" and those below 1% "highly significant." However, there is no solid mathematical reasoning for these choices. There may be circumstances under which a p value of 7% is a low enough probability to cast doubt on the assumed hypothesis of no effect; there may be other circumstances where a p value needs to be considerably below 1% before a reasonable scientist will accept the data as evidence against the hypothesis of no effect. The working toxicologist should arrive at a trial with some prior idea of what effects are expected. If there is an apparent effect due to treatment that runs contrary to expectation, the

toxicologist may want to resist overthrowing this prior knowledge on the basis of a mere 5% p value. Likewise, failure to
find formal significance when there appears to be an overwhelm-
ing difference should not be a reason for ignoring it.

Directing Statistical Tests

In the previous sections we examined the use of ANOVA tests
to compare the scatter of data among different treatments.
Both Fisher's Gaussian ANOVA and the Kruskal—Wallis tests are
general procedures that compare the observed data against all
possible patterns different from the *null hypothesis* that there
was no treatment-related effect. However, a fundamental principle of toxicology is that most toxicity is dose related. We
can increase the power of the statistical procedure (its ability
to detect a treatment effect when there is one) if we can limit
the analysis to one that considers the null hypothesis against
only the possibility of a dose-related increase in toxicity. Then
certain results may become significant when tested in this restricted way, although they did not produce low p values in a
more general ANOVA. Suppose we observed the following average changes in SGOT levels:

Dose administered mg/kg	Average change in SGOT
0	−2.5
10	3.4
50	12.5
100	22.2

Such a pattern suggests the type of dose response that seems
reasonable in the light of toxicology. If we apply a general
ANOVA, we may not get a significant p value. If, however, we
apply restricted tests that take into account the possibility of
a dose response, we stand a better chance of detecting an effect when there is one.

One such test is *Bartholomew's test of monotone trend* (Barlow et al., 1972). In Bartholomew's test the data are fit to the
hypothesis of no effect by comparing the individual treatments
values to the average of all the values. The data are then fit
to the hypothesis that there is some sort of a dose response.

We compare the two goodness-of-fit statistics to create a statistical test directed against a dose response. One problem with Bartholomew's test is that it assumes a Gaussian distribution to the data and the computer algorithms for calculating p values are complicated. As a result, most of the widely available statistical computer programs are not able to compute Bartholomew's test.

Bartholomew's test is the most general one that is directed against dose response alternatives. Within the framework of ANOVA there exists a more restricted test called the test for *linear contrasts in dose* (Dixon and Massey, 1969). This compares the fit of the data to a single overall average (the null hypothesis) against the fit of the data to a linear dose response. When the dose response is not linear, this test might still detect an effect, as long as the dose response tends to increase with dose. It is less powerful against a "dose response" where the lower doses have no effect and all the effect is found only in the highest dose.

When the F test from ANOVA (or the chi-squared test from Kruskal—Wallis) is used as the test of significance, the null hypothesis of no effect is compared to all possible patterns of effect. Since many such patterns are possible, the F test lacks power. However, even if we do not restrict the alternative to a dose response, other reasonable restrictions can be applied to the ANOVA that lead to more powerful tests.

1. *Dunnett's t test* (Dunnett, 1955) assumes that there is one treatment called the control and the only comparisons that will be made will be between the control treatment and each of the others. If one does not intend to run statistical tests comparing active treatments, Dunnett's t test is appropriate and powerful.

2. *Newman—Kuells* (Dixon and Massey, 1969) contrasts are constructed on the assumption that the treatments can be ranked (e.g., in terms of dose) and the only comparisons that will be made will be between a given treatment and the next highest-ranking one. The procedure then tells us where the significant gaps are between effects. Suppose we had a response pattern similar to the one described just previously, with mean changes in the SGOT level of -2.5, 3.4, 12.5, and 22.2. The Newman—Kuells contrasts tell us whether the break from -2.5 to 3.4, 3.4 to 12.5, or 12.5 to 22.2 is significant.

3. The Newman—Kuells contrasts are examples of a more general method known as *orthogonal contrasts* (Dixon and Massey, 1969). The idea here is to limit the number of possible comparsons to those that will allow the use of more powerful test sta-

tistics. One frequently useful orthogonal contrast compares the data associated with a given dose with the average data of all the lower doses. Thus, in the previous example, we would compare -2.5 to 3.4, the average of -2.5 and 3.4 to 12.5, and finally the average of -2.5, 3.4, and 12.5 to 22.2.

The use of Dunnett's t enables one to determine which treatments differ from the control. The use of Newman—Kuells or the other orthogonal contrasts enables one to determine where the dose response increases. Overall tests of determining whether or not a dose response exists can be run with Bartholomew's test or with a linear contrast in dose.

Although it is widespread practice to apply only such general tests as ANOVA to toxicological data, statistical theory suggests that it would be more sensible to apply these more powerful tests. However, the rules of statistical theory that are involved can become quite complicated. Some practices that might seem reasonable to the toxicologist will produce erroneous results. One practice (also widespread) that will produce false p values is to ignore the fact that several doses are being run. The toxicologist might be tempted to carry out one statistical test to compare just the controls and a given dose, another test to compare the controls to another dose, another one to compare one dose to the next, and so on, each test ignoring the existence of the other doses. Each of these hypothesis tests has a probability of producing a false positive. In general, if a test is run and a p value of 5% is called "significant," there is 1 chance in 20 (5%) that this test will produce a false positive. This is a reasonably small chance for most scientific work; however, if many tests are run on the same experiment, each one with a 5% probability of finding a false positive result, the chances of obtaining at least one false positive in that experiment increases. Monte Carlo studies run on computers have shown that the chance of a false positive can rise as high as 50% under some circumstances. Dunnett's t, Newman—Kuells, and the use of linear and other orthogonal contrasts all run the hypothesis tests within a single larger framework, guaranteeing that the chance of at least one false positive finding in the entire analysis is kept at the appropriate p level chosen to indicate significance.

Confidence Bounds on Comparisons

Closely related to the idea of significance testing is the concept of *confidence bounds*. We already encountered this concept in

Unit I on the LD50. There the best-fitting sigmoid curve was bounded above and below by other curves and it was possible to read off bounds on the LD50. Confidence bounds can be constructed under other circumstances.

Suppose that in the example of changes in SGOT levels we suspect that the highest treatment dose caused an increase in SGOT. Since the study ran for several months, the control group also had a change in SGOT levels. If the treatment group experienced an effect, their change in SGOT levels would have to have been greater than the "normal" change seen in the controls. At this point we conceive of some ideal experiment involving millions of animals—so many animals, in fact, that the average change in SGOT level among the controls would average out all random noise and represent the true normal change. Similarly, in this ideal experiment the average change in SGOT level in the highest dose group represents the true effect of that dose.

Compared to this ideal experiment, the one we can conduct has so few animals that the average change in SGOT level for one of our experimental groups represents the true effect plus or minus some unknown random number. We can use a hypothsis test to check whether the observed average change differed significantly from zero or whether it differed from 10 or any other number we wish to name. In fact, we can run through an entire sequence of numbers, finally finding that difference which just barely produces a significant p value. If zero cannot be rejected, then we can find some number greater than zero that is just barely rejected and some number less than zero that is just barely rejected. These two numbers form confidence bounds on the true underlying mean change that might have been seen in a similar, very large study.

The *coverage* of the confidence bounds depends upon what p value was taken to indicate significance. If a p value of 5% was chosen, then the coverage is 95%; if the significant p value is 1%, then the coverage is 99%.

The mathematics are quite complicated, but this basic idea of trying a sequence of values until finding those that are just barely rejected can be applied to the difference in mean effect between any two treatments. If this is done within the context of a protected and restricted statistical test, like the Newman—Kuels contrasts or Dunnett's t, one can be reasonably sure that the coverage is as indicated. If one tries to construct confidence bounds based upon several different statistical tests comparing different pairs of treatments, without adjusting for the multiplicity of tests, then the computed confidence bounds will not have the indicated coverage.

Thus most computer programs for computing ANOVAs (Fisher's Gaussian ANOVA or the Kruskal—Wallis nonparametric ANOVA) contain a provision for computing confidence bounds on any comparison of treatment effects. This is sometimes referred to as a *contrast*. A contrast is a weighted sum of the treatment effects in which the weights add up to zero (some are negative and some are positive.) If the computer program has a provision for computing confidence bounds in an interactive mode, the computer will ask the user for the weights to be used in the contrast desired. The most common contrasts are simple comparisons between two treatments. If there are four treatments (controls plus three dosing groups), a contrast comparing the first to the second would have the weights

$$1, -1, 0, 0$$

A contrast comparing the first treatment to the third would have weights

$$1, 0, -1, 0$$

and so forth. If the toxicologist wished to compare the average effect of all the treatments versus the controls, he could use the weights

$$1, -1/3, -1/3, -1/3$$

Technically, there are a number of different methods for calculating confidence bounds on contrasts from an ANOVA, and the computer program may distinguish between Sheffé and Tukey contrasts, for instance. However, almost all of these methods produce bounds that are close to one another.

One method for computing an approximate confidence bound that appears often in the literature is the method of reporting average effects with a standard error or standard deviation in the form

$$a \pm s,$$

where a is the observed average and s is either the estimated standard deviation of the data or the estimated standard error

of the mean. If s is the standard error of the mean, approximate 95% confidence bounds on the mean can be computed by
adding or subtracting two standard errors from the observed
average. However, this method does not take into account the
existence of several treatments in the experiment, and attempts
to compute the standard error of the difference in mean effects
from this sort of display can lead to errors. To be reasonably
sure of the coverage, it is recommended that the toxicologist
use an ANOVA computer program that computes the confidence
bounds on the contrasts to be investigated and that the average
difference in effects be reported along with the appropriate confidence bounds.

Other Observations on Living Animals

In the previous sections we treated the statistical tests as if
only one measurement were made on the living animals (e.g.,
SGOT). In fact, blood chemistries describe as many as 20 different enzymes. Along with hematological measures, a subchronic toxicity study will also accumulate ophthamological data
and daily observations of signs, such as abnormal gnawing behavior or extreme lassitude. It has become standard to subject
the weights, clinical chemistries, and hematologies to statistical
analysis and to report other observations without analysis, probably because these first measures lend themselves to standard
statistical procedures. But the toxicologist should be aware
that although there are no standard methods of subjecting these
other measures to formal statistical review and, as a result,
they form only a small part of the final toxicological report,
there may be times when they are far more important to the interpretation of the study than all the formal statistical tests applied to continuous measures.

One way to handle these other observations is by counting
events. The toxicologist could consider as a "measure" of the
effect of treatment the number of days an animal is observed to
display a given sign or the degree of the sign might be expressed on a three- or four-point scale (as is often done for emesis
in dogs). Once the observations have been formalized and reduced to numbers, these numbers (one for each animal) can be
analyzed by the methods described earlier.

The way in which the observed sign is reduced to a number
can affect a statistical analysis's sensitivity in detecting a treatment-related effect. This can be thought of in terms of the

amount of information supplied by each animal. The greatest
amount of information is given when the number associated with
an animal comes from an infinitely dense scale of numbers (such
as the measurement of SGOT levels). If the animals are categorized into four groups, such as those with "no effect," those
with "slight effects," those with "moderate effects," and those
with "severe effects," for instance, less information is derived
from each animal. The least amount of information is obtained
when a single animal is simply categorized as having been affected or not.

We can think of statistical analyses as using the available
information to detect treatment-related effects. If little information is obtained per animal, a greater number of animals will
be needed to supply the same amount of total information as
when a large amount of information is obtained per animal. Thus
whenever data are reduced to counts of animals with an effect,
the toxicologist's chances of detecting whether that effect is
treatment related are usually reduced.

Observations Made on Organ Weights
at the Time of Death

For the toxicologist the observations made at necropsy are often
the most important in the interpretation of the study. They
are also the area where standard statistical procedures are the
least developed, probably because it is not obvious how to reduce such items as cause of death, gross pathology, and histopathology to measurable quantities.

However, one aspect of gross pathology lends itself to statistical analysis: the weighing of the major organs. The toxic
effects of chemicals often atrophy certain organs or enlarge
others, so it becomes useful to compare the individual animal's
organ weights across treatments. As with the continuous measures of blood chemistries, these numbers are easily handled by
ANOVA techniques, and significant differences due to treatment
can be detected by the appropriate use of dose response tests
or contrasts.

Toxicologists disagree as to whether absolute organ weights
or relative organ weights (the organ's weight divided by the
total animal weight) should be compared. If, for instance, the
treatment chemical gave a bad taste to the food, causing the
treated animals to eat less and in turn lowering their final
weights, it is argued that many organs will have a lower abso-

lute weight but the same relative weight. On the other side of
the argument, some believe that certain organs (such as the
brain) are independent of the animal's overall weight. A com-
promise is often reached by displaying analyses of both types
of measures in the toxicological report.

There is a statistical method of getting around this issue
called the *analysis of covariance* (Dixon and Massey, 1969). It
was initially developed by Fisher in agricultural research. It
had been proposed that the use of some fertilizers increased a
plant's size, but with most of the growth going to staw and
very little to the grain that was to be harvested. Fisher suggested using the amount of straw as a "covariate," as something
else affected by the treatment, whose component had to be removed from the analysis. The concept of a covariate is relatively vague. In order to use it in a formal statistical analysis, one has to define it in terms of a specific mathematical
formula. Fisher's solution is the one most widely used today.
It proposes that the measure in which one is interested (e.g.,
the organ weight) be designated by the letter Y. The covariate is designated by the letter X. The relationship between
X, Y, and the effect one wishes to measure can be written as

$$Y = a + bX$$

where a stands for the "true effect" one is interested in measuring and b represents the relative influence of the covariate
on the observed effect. The statistical techniques of analysis
of covariance allow one to test for differences in the true effect a between treatments just as one is able to do with analysis
of variance, using directed tests for dose response or contrasts.

The astute student will realize that Fisher's model is not a
description of what most toxicologists think of when they propose using relative organ weights. These toxicologists usually
have in mind a model such as

$$Y = bX \text{ or } \frac{Y}{X} = b$$

where b represents the "true" measure. This is a much more
restricted mathematical model than Fisher's. It assumes that
the ratio of the organ weight to the total animal weight describes the entire relationship between the two variables. Fisher's model says that there is some underlying "true" weight to

or from which the animal's general pattern of growth has added or subtracted something (b can be negative). If the organ in question has a weight independent of the total animal weight, the information in the data will cause Fisher's model to estimate b as close to zero.

Since Fisher's model is more flexible, it can cover both the case where the total animal weight has a major influence and the case where it does not. Furthermore, it can tell whether the treatment had an effect on the true weight a or the relative effect b. Thus a computer program that runs an analysis of covariance on organ weights will print out more than just a comparison of treatments. It will print out a test of whether the relative effect b is constant across groups, as well as the confidence bounds on that relative effect. Most computer programs for analysis of covariance will also allow for a more complicated, possibly a quadratic or cubic, function of the total weight X.

The disadvantage of a complex model lies in its very flexibility. If, by using various tests of significance, the computer program decides that an analysis using a complicated function of X, like a cubic equation, is the best-fitting one, the toxicologist is forced to interpret the meaning of that complicated function. In particular, if the *test for parallelism*—the test of whether the same function fits for all treatments—is ignored, the toxicologist can come to conclusions that may make little or no biological sense.

Other Observations Made at the Time of Death

At time of necropsy observations of overt organ toxicity are also made. Some of these observations (such as nephritis) can be graded, while others can be scored only as occurrence or nonoccurrence. Questions arise about how to treat animals that died early. All of these problems (and others) occur in the analysis of chronic toxicity studies and are discussed in the next section.

The Overall Aim of the Study

The first section of this chapter quoted FDA guidelines to show that the overall aim of a subchronic toxicity study is twofold: to establish a starting dose for another study—"safe dose" for human trials or "maximum tolerated dose" for chronic studies—

and to describe the nature of the toxicity. The statistical techniques discussed in this chapter are merely methods for organizing the large amount of data accumulated in such a study. None of them provide direct solutions to these problems. At some point the toxicologist must sit back with the tables of significance tests and confidence intervals, his knowledge of the putative pharmacology of the test compound, and his prior experience with patterns of toxicity of previous studies and decide for himself what are usable solutions to these problems. Can statistical methods be used at that point?

The answer is qualified yes.

All of the procedures described earlier in this chapter deal with *univariate data*. The process of integrating all the data from a study into a final answer requires that one consider the interrelationships among the measurements. Significant elevations in the SGOT level, for instance, unaccompanied by elevations in other liver enzymes and with no indication of liver damage at necropsy must have a different interpretation than such a finding accompanied by elevations of serum glutamic-pyruvic transaminase and serious liver pathology. One way of using statistical procedures to aid in this interpretation is to examine sets of variables that have some biological connection. These are the methods of *multivariate analysis* which are widely used in sociology and psychology but seldom seen in toxicological journals.

Since multivariate procedures are well developed in statistics but there is no consensus as to how they might be applied to toxicology, this section will only sketch what these techniques can do and leave it to the student who wishes to use them to seek out a knowledgable statistician for more details.

Liver toxicity, as an example, can be reflected in the measurements of four to five blood chemistries, the weight of the organ on necropsy, and cellular changes seen in histopathology. To use statistics, we must reduce all of these to numerical measures. Suppose, however, that we can grade the liver histopathology on a scale from 1 (normal) to 5 (severe damage); we can then add this measure to the other five or six. If several types of liver damage are seen and if each of them can be graded, these measures can be added to the vector of other measures.

The resulting set of 6—10 numbers derived from each animal can be visualized as a point in, say, 10-dimensional space. No one can really visualize 10-dimensional space, but most of us think of it in terms of three dimensions. Each animal in a group will supply a single point in this space, so the scatter of points

will tend to gather in a ball about a center that can be determined by plotting the average values for all the animals. Just
as in the Kruskal—Wallis ANOVA, we can think of this scatter
of points as overlaying that from another treatment group. If
the two balls of scatter are completely separate, this provides
fairly strong evidence that there is a difference in effect due to
treatment. As might be expected, there is a statistical technique
for analyzing the different patterns of scatter and it is known
as *multivariate analysis of variance (MANOVA)*.

In the discussion of univariate analysis of variance, we noted
the choices the toxicologist had in methods of analysis and their
interpretation. Expanding the problem into 10 dimensions increases this complexity of choice. There are now several test
statistics (not just a simple F test), each of which might produce
a different degree of significance from the others; the methodology rests more heavily on assumptions about the underlying
probability distributions; and there is a wider variety of methodology to use.

Another area where statistical techniques might be useful involves the inclusion of information external to the study itself.
If, for instance, 3 out of 50 of the high-dose animals show a
very rare lesion, a formal statistical test comparing 3 of 50
versus 0 of 50 for the controls will not be "significant" by itself; however, the fact that the toxicologist has never seen
this lesion in control animals for the past 10 years of similar
studies should mean something. Or suppose there is a lesion
whose incidence increases with dose and where the test of dose
response is "nonsignificant." But, the lesion is an obvious
consequence of the known pharmacology of the test compound.
In both these cases, use of external information leads to a much
stronger conclusions than the statistical tests will allow.

Statistical techniques which take external information into
account are called *Bayesian statistics* after Thomas Bayes, who
discovered a formula in the eighteenth century that enabled him
to determine the probability of an event that had already occurred with a knowledge of the events that followed it. Bayesian techniques form the basis of well over half of the theoretical
articles that appear in journals such as the *Annals of Mathematical Statistics* but are very seldom used in scientific practice.
One reason for this is that they usually require very arbitrary
assumptions which cannot be verified through the data.

Thus the qualified yes to the question posed earlier. To follow through on an earlier analogy, statistical techniques are like

a glass into which you pour wine. The glass may aid you in
examining the wine's clarity, but the taste and bouquet you
must judge for yourself, independent of the glass.

5

Lessons from Subchronic Studies

Well-Defined Questions

Toxicology is an example of science used to answer vague questions. The fundamental ones examined in subchronic tests are

1. What is a "safe" dose?
2. What is the nature of the induced toxicity?

For drugs intended for human use, the answer to the first question depends upon determining where pharmacological activity ends and where toxicity begins; for environmental contaminants, this depends upon the nature of the exposure. The answer to the second question is qualitative and descriptive.

On the other hand, statistical techniques can be used to answer only well-defined questions. These are questions built around the exact values of parameters in formal mathematical models. A statistical technique like the ANOVA can be used to determine whether there is an increase in mean SGOT change that is linearly related to dose. Another technique can be used to determine whether there is an increase in the proportion of animals with severe SGOT changes that is related to dose. These are two distinctly different mathematical questions, although their answers may play a role in determining the qualitative answer to a single vague toxicological question.

Thus the first lesson to be learned from examining the statistical techniques used in subchronic toxicity tests is the following:

1. The statistical analyses in these studies cannot be used as if they produced immutable, definitive answers. They must be interpreted to determine how closely the well-defined mathematical question matches the vague toxicological question.

Looking for the Car Keys

The major thrust of statistical analysis in subchronic studies deals with things that are clearly measured in terms of numbers—blood chemistries, organ weights, weights of growing animals, and so on. Because so many such numbers are generated, this results in a large collection of statistical hypothesis tests and confidence bounds. About 90% of the pages in a complete report of such a study consists of tabulations of the data and the associated statistical analysis. The reader cannot help but conclude that most of these analyses were done because the data were available and easily analyzed. It is like the tale of the drunk who was searching for his missing car keys under the streetlamp because the light was better there. Findings of statistical significances or nonsignificances here may or may not be indicative of the toxicity of the test compound. They must be interpreted in terms of the final question being asked.

The large number of analyses run on these data provide more problems than possible irrelevancy. Each one deals with only one small aspect of the study. (Did the SGOT change from week 0 to week 4?) As more and more of these analyses are run, the probability that purely spurious events will be tagged "significant" increases. This is the problem of multiple hypothesis testing. Furthermore, as the number of analyses increases, it is harder for the toxicologist to see the whole picture at once. As a result, statistical techniques, which should be a means of reducing a large amount of data to the essential bits of information, become a source of further confusion.

Thus we can cite two more lessons, the third one having already been noted in Chapter 4:

2. Every number need not be analyzed simply because "it is there." If statistical analyses are to used to reduce data to information, they must be used with discretion.

3. It may be useful to apply the techniques of multivariate and Bayesian statistics to organize the information in a more useful way.

What Is "Significant"?

Findings of statistical significance are often used as flags to identify toxic events in a study. As a first cut through the data, this is often a useful procedure. If statistical techniques were not used, the toxicologist would be forced to examine the average responses for each treatment and come to some conclusion about whether there was a treatment-related effect. It has been this author's experience that statistical techniques usually flag events as significant that would have been missed by such "eyeballing" of the data. Thus hypothesis tests are an effective means of identifying measures for which there might be a treatment effect.

However, the choice of the p value that is associated with significance is purely arbitrary. A p value of 6.3% is not less useful than a p value of 4.7%. Since the statistical analyses use approximate procedures and since we can never be really sure of what the true probability is, we have to view these as two numbers which represent the same degree of improbability. Yet the unfortunate use of a word like *significance*, which now contains more implicit meaning than Karl Pearson originally intended, leads us to stare hard at p values of 4.7% and ignore those of 6.3%.

The lessons here are the following:

4. Statistical "significance" should be interpreted in the Pearsonian sense, as indicating that something has happened but not necessarily something important.

Since the concept of significance testing deals with the improbability of the observation under the null hypothesis tested, it cannot be used as a tool for accepting a null hypothesis but only as a tool for rejecting. Thus we can say the following:

5. A failure to find significance does not necessarily mean that there was no treatment effect.

However, it is possible to increase the sensitivity of a statistical test for detecting a treatment effect. This is done by directing the test, cutting down on the range of possible alternatives, and including only those that might make biological sense. The ability of a statistical test procedure to detect a difference when one actually exists is called the *power* of the test. There are a number of ways in which the power of a test can be increased. One of these is through the use of directed tests. This leads to a sixth lesson:

6. If the biological knowledge is sufficient to propose a restricted set of alternatives (such as a dose-related increase), then it is possible to run a more powerful statistical test to determine if there is a treatment effect.

6

Theory Behind Statistical Tests

Kruskal—Wallis Nonparametric ANOVA

Suppose we had 10 animals randomly assigned to five groups. If each animal has a number from 1 to 10 assigned to it, one such configuration might be as follows:

Group	Animals assigned
A	4, 10
B	1, 2
C	7, 9
D	6, 8
E	3, 5

If the numbers are assigned at random, each group should have an average number not too far from 4.5 (the average of all the numbers from 1 to 10). The sum of the squared differences between the average of each group and this theoretical average of 4.5 will measure how far the pattern falls from random expectation. There are almost 1000 permutations of the way in which these 10 numbers can be assigned to five groups. For each of these permutations we can compute the sum of the squared deviations; for some of them the resulting sum will be the same as for others. If the five groups are assigned at random, then each of the possible permutations is equally probable, and we can compute the probability distribution of these sums.

If, however, the groups are not assigned at random, but so that groups A and B have a higher probability of getting lower numbers and groups D and E a higher probability of getting numbers near 10, then the probability distribution of the sum of the squared deviations will be different. In fact, it will be such that large values are more probable than small ones, since four of the groups will have a high probability of averages far from 4.5.

Suppose we wished to determine whether the assignment was completely at random or due to a weighted scheme that tended to make groups A and B have low numbers and groups D and E high numbers. It would seem reasonable to examine the sum of squared deviations of the data observed and reject the first hypothesis (equal probabilities) in favor of the second one if the sum is very large.

This is the idea behind hypothesis testing in general and the type of computation behind the Kruskal—Wallis test in particular. Suppose, for instance, that we observe changes in SGOT levels among four groups of 50 animals each. If we rank all 200 values in size, we can replace the observed actual values by their ranks. We have thus converted the data into a set of numbers from 1 to 200. If the treatments had nothing to do with the changes in SGOT, the set of ranks from 1 to 200 will appear to be randomly assigned with respect to treatment groups. As with the numbers from 1 to 10 assigned to five groups, we can compute the probability distribution of the sum of squared deviations from the random average of 100.5 under this hypothesis of random assignment. If the assignment was not random but due to some influence of treatment on changes in SGOT, then the sum of squared deviations will tend to be larger in value.

Thus we can reject the null hypothesis of random assignment
if the sum of squared deviations is improbably large.   The
Kruskal—Wallis test statistic is this sum of squared deviations.
The p value produced by a computer program is the probability
of a value being as large or larger than the one observed, if it
is due to purely random assignment.

Fisher's ANOVA

The concept of squared deviations from an overall mean that
lies behind the Kruskal—Wallis procedure also plays a role in
Fisher's earlier development of Gaussian analysis of variance.
For Fisher's derivation, we assume that the five sets of numbers come from five Gaussian distributions.   The Gaussian distribution is fixed by two parameters, the underlying mean and
the underlying variance.   If all five of the distributions have
the same variance, then they would differ, at most, in their
means.   So, following the same type of reasoning as in the
Kruskal—Wallis derivation, we would look at the sum of squared
deviations of the averages of each group from the overall average of all five groups taken together.   If this sum of deviations
is large, the underlying means differ.

Unfortunately, the variance that is common to all five distributions is unknown.   If it is small, then the sum of the
squared deviations need not be very large to imply differences
in the underlying means.   If the variance is large, then a much
larger sum of squared deviations is needed to imply differing
means.   Fisher's solution was to consider the following ratio:

Sum of squared deviations/estimate of the variance

This ratio (with adjustments to account for the number of observations involved) is the F test of a Gaussian analysis of
variance.   Starting with the assumption that the original data
were Gaussian with a common variance, it is possible to compute
the probability distribution of this ratio under the assumption
that the means are all the same (or under any other specific assumption about the relationships among the means, for that matter).   Values of the computed F test that are improbably large
under the distribution implied by a common mean are taken as
evidence against that null hypothesis.

The null hypothesis tested in the Kruskal—Wallis test is that
the five sets of numbers all came from the same distribution

(with no restrictions on what that distribution might be). The null hypothesis tested in the Gaussian analysis of variance is that the five sets of numbers all came from the same Gaussian distribution. What happens, then, if the true underlying distribution is not Gaussian or if it is Gaussian with a common mean but differing variances? This is the question of statistical *robustness*. How accurate is the computed p value if some of the underlying assumptions are not true?

It can be seen that this is a vague question. What do we mean by "some" of the assumptions? If there is no common underlying variance, how far different can the variances be? If the underlying distribution is not Gaussian, is there some larger class of nice distributions that includes the Gaussian for which the p values will be correct? All of these are theoretical questions that have not been answered completely. A large number of computer simulations (Monte Carlo studies) have been run on what happens to the true p value under differing violations of the assumptions. The general consensus of these studies is that the p values tend to be correct or slightly too large as long as the underlying probability distribution is symmetrical and the variances do not differ by more than threefold. Thus it is the widely held opinion among statisticians that this type of analysis of variance is "robust" and can be used in most situations.

Neyman—Pearson Formulation of Hypothesis Tests

The previous discussions dealt entirely with the distribution of the test statistic under the null hypothesis being tested. It is possible to think of hypothesis tests only in terms of the null hypothesis, constructing various ways of testing that null hypothesis and applying them as one sees fit. However, if the null hypothesis is not true, then there has to be some probability distribution that is true, and the nature of that true probability distribution should dictate the type of test to be used.

For instance, suppose we are examining values of SGOT at week 4 of a study. The SGOT has one interesting characteristic to its probability distribution that has often been noted. If you have two sets of such data, the one with the smaller average value tends to have a smaller variance. In fact, the variance increases almost as fast as the square of the average. Thus, if there is a treatment-induced increase in SGOT, the higher doses will show not only an increase in mean but also an increase in variance. This can be seen in the data displayed in Table 1. Although the two highest doses show much larger values of SGOT than the lower doses or the controls, it can be seen that the increased spread inserts some of those values among the scatter of numbers in the control group. The Kruskal—Wallis test could be fooled by such a result, failing to find significance because there are too many high dose values scattered among the low dose ranges.

If, on the other hand, we know that a treatment-related effect will cause an increase in both the mean and the variance, we could construct a statistical test sensitive to that kind of an alternative hypothesis. It was this type of thinking that lead to a set of five collaborative papers between J. Neyman and E. S. Pearson (the son of Karl Pearson) in the late 1920s (Neyman and Pearson, 1933; Neyman, 1935). These papers proposed a formal mathematical structure for hypothesis tests that has been widely accepted and which is now taught in almost all elementary statistics courses. It is not the only way of viewing hypothesis tests and it engendered a great deal of controversy when first proposed; in fact, there are prominent schools of statistical thought that continue to reject this formulation. It limits very severely the range of application of hypothesis tests, and the practicing scientist should always be alert to whether this restricted formulation is really appropriate to the current investigation.

The Neyman—Pearson formulation assumes that we have a well-defined null hypothesis that we wish to test and that there is an equally well-defined class of alternative hypotheses. We assume that, if the null hypothesis is not true, then the true probability distribution of the data we have at hand can be described by one of the class of alternative hypotheses. For instance, if we wish to use the Kruskal—Wallis test, the null hypothesis is that all groups of data came from the same probability distribution. However, it is of little use to propose that the set of alternatives contains all other possible distributions.

This can be seen by considering an alternative where all the groups have the same mean but the variance increases with increasing dose. The Kruskal—Wallis test would be blind to such an alternative and be unable to reject the null hypothesis. In fact, the class of alternatives for which the Kruskal—Wallis test is the most appropriate is very narrow. It consists of the set of distributions where the plot of the cumulative probability for one dosing group lies entirely below that of a lower dose, as in

Table 1 SGOT Values at Week 4

Controls	Dose 1	Dose 2	Dose 3
123	170	157	181
153	141	181	311
125	160	189	250
120	152	197	337
130	152	153	306
127	158	177	291
142	142	198	319
120	145	180	278
119	165	169	365
142	167	151	261

Figure 1. If the alternative hypothesis is such that the cumulative probability curves cross at any point, there are better
tests than the Kruskal—Wallis.

What do we mean by *better*? To answer this, Neyman and
Pearson considered a simple case where the null hypothesis is a
single well-defined probability distribution (such as all five
groups come from a Gaussian distribution with mean 30.6 and .
variance 952.7) and where the alternative hypothesis is also a
single well-defined probability distribution (the controls and the
low-dose group have a Gaussian distribution with mean 30.6 and
variance 952.7, the middle-dose group has a Gaussian distribution with mean 44.9 and variance 1055.3, and the high-dose
group has a Gaussian distribution with mean 86.2 and variance
2967.5). This is, of course, a situation that can never occur,
we can never know two exact distributions and be absolutely
sure that the data come from either one or the other. But it is
a useful trick in theoretical mathematics to start with a conceptually simple case and build from there.

If we had such a simple situation, then there are two types
of errors we might make. We could decide that the alternative
hypothesis is true when, in fact, the null is the true one.
This is called a *type I error*. Or we could decide that the null

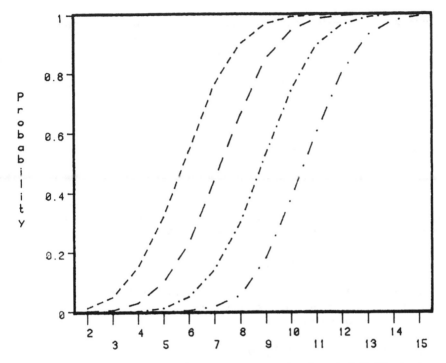

Figure 1   Pattern of distributions where the Droskal—Wallis test is most powerful.

hypothesis is true when, in fact, the alternative is true.   This is called a *type II error*.   The situation is usually diagrammed as in Table 2.   Let us suppose we set up a rule for deciding which hypothesis is true.   For example, we might compute the Kruskal—Wallis test statistic and decide in advance to reject the null hypothesis and accept the alternative if the computed p value is 5% or less and to accept the null hypothesis otherwise.

For any decision rule we pick, the random fall of the data will induce some error into our decision.   We can, however, determine the probability of a type I error by computing the probability of our decision rule rejecting the null hypothesis using the distribution of the null hypothesis.   Similarily, we can compute the probability of a type II error from the distribution of the alternative hypothesis.   We call the probability of a type I error the *alpha level* and the probability of a type II error

Table 2 Neyman–Pearson Formulation of Hypothesis Testing

	True state of nature	
Decision	Null hypothesis	Alternative hypothesis
Null hypothesis	No error	Type II error
Alternative hypothesis	Type I error	No error

the *beta level*. It can be seen that there is an inverse relationship between alpha and beta. At one extreme we can drive the alpha level to zero by using the decision rule

Never reject the null hypothesis.

Whenever the alternative hypothesis is true, this decision rule will produce an error, so the resulting beta level is 100%. Similarly, if we adopt the rule

Always reject the null hypothesis

the beta level will be zero but alpha will equal 100%. Figure 2 displays the general inverse relationship between these two probabilities of error.

Neyman and Pearson showed that there is no "best" solution to this situation, because of the inverse relationship between the alpha and beta levels. However, they were able to show that there is a single best decision rule that will make the beta level smallest for any fixed alpha level. That is, if one decides in advance what probability of a type I error one can live with, there is a "best" statistical test that will give the lowest possible beta level for the alpha chosen. This is called a *most powerful test*.

Neyman–Pearson Hypothesis Testing

The situation described earlier of a simple null hypothesis and a simple alternative is only a mathematical fiction, designed to fix ideas and examine them in a conceptually simple situation.

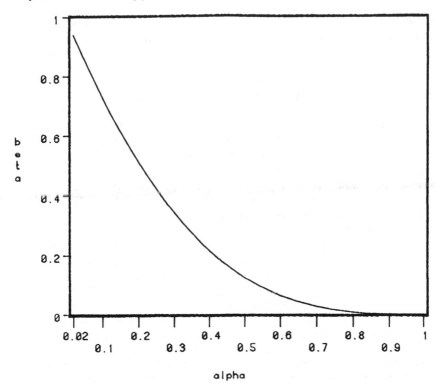

Figure 2 Probability of a type I error versus the probability of a type II error.

Neyman and Pearson then considered a slightly more realistic situation. Let us assume we have a well-defined simple null hypothesis as before but we let the alternative hypothesis come from a class of probability distributions. For instance, we can let the null hypothesis be that all five sets of data come from a single Gaussian distribution with a mean of 32.5 and a variance of 66.4 but let the alternative hypothesis come from the class of distributions where all five groups have a variance of 66.4 but at least two of the means are greater than 45.0.

We can now use fairly complicated mathematics to compute the beta level associated with any specific member of the class of alternatives for any decision rule we might choose. Furthermore, it is sometimes possible to find a decision rule that is uniformly more powerful than any other, that is, one which

produces a lower beta level for every member of the class of
alternatives than any other decision rule. In the situation described above the most powerful decision rule is one that uses
the F test from Fisher's analysis of variance.

By carefully expanding the class of alternative hypotheses,
Neyman and Pearson were able to establish situations where
there was a uniformly most powerful procedure and situations
where no such procedure could possibly exist. As the student
might surmise, the situations where there are uniformly most
powerful tests are very restricted ones. For instance, suppose
we wish to determine if the underlying mean change in SGOT
levels from week 12 to week 14 in the high-dose group is different from zero, assuming that such a change might be due to
treatment. If we let the alternative hypothesis be any change
(increase or decrease), there is no uniformly most powerful
test. If, however, we can assume that any change would have
to involve an increase, then such a test exists and can be used.

The major lesson here for the toxicologist is that it pays to
restrict the class of alternative hypotheses by careful consideration beforehand to what might be expected if the treatment induces toxicity. This is the idea behind using the dose response
as a defining characteristic of the alternative hypotheses. In
the late 1950s Neyman returned to the question of constructing
uniformly most powerful tests for restricted classes of alternatives and worked out a general procedure known as the *restricted chi-squared test* (Fix et al., 1959). The tests of dose response described in Chapter 4 are examples of Neyman's restricted chi-squared tests.

There are two flaws in the Neyman--Pearson formulation of
hypothesis testing. The first arises when we try to expand
the null hypothesis from a single well-defined distribution to a
class of distributions. Suppose, for instance, that we examine
the changes in SGOT level in the high-dose group from week 12
to week 16 and wish to test the null hypothesis that the underlying mean change is less than 10 versus an alternative that it
is greater than 10. (Now we have an infinite number of null
hypotheses, since the mean can be any number less than 10 and
still be part of the null hypothesis.) There is no statistical
test that will guarantee the alpha level for all the members of
this null hypothesis.

We can construct a uniformly most powerful test that examines the null hypothesis that the mean change is equal to 10,
and we can set the alpha level for that test at, say, 5%. However, if the true null hypothesis is that the mean change equals

5 units, then the probability of a type I error is less than 5%.

The situation described above is not unreasonable. By expanding the null hypothesis to a class and testing as if the most extreme member of that class were the null hypothesis to be tested, we have, at worst, reduced the probability of error. However, unless we are careful, we can construct what appears to be a sensible hypothesis test for a class of null hypotheses where the probability of a type I error is, in fact, greater than the proposed alpha level for some members of that class. Such a situation occurs when examining a large number of possible lesions and testing if any one of them implies "significant" toxicity for the treatment being tested. It is an apparent paradox of statistics that the rule which says that significance is declared if any of the individual tests is significant is one that produces a probability of a type I error greater than alpha for the situation where no toxicity is associated with any of the lesions.

The flaw discussed above deals with the problem of statistical false positives (probability of a type I error). The other flaw of the Neyman–Pearson formulation deals with the problem of statistical false negatives (probability of a type II error). Neyman and Pearson recognized early on that it is impossible to find a uniformly most powerful test for many realistic situations. What was not realized until the late 1950s, in an article due to Bahadur and Savage (1956), is that it is impossible to find a test with any power at all unless the class of alternatives is restricted. Bahadur and Savage proved that if the class of all possible distributions is considered as the alternative, then there exist some alternative distributions for which the probability of a type II error is as large as it can possibly be (the complement of alpha).

It is not clear whether the types of distributions discussed by Bahadur and Savage can occur in real-life problems, but, if they do, it will be in a situation where some animals are predispossed to a particular lesion and others are resistant to that lesion and where the treatment modifies the pattern of lesions for one type of animal but not for the other.

Thus the Neyman–Pearson formulation of hypothesis testing provides a widely used basis for statistical analysis. It has a reasonable structure and involves some useful ideas, such as power. Lacking anything better, it is a good conceptual framework for viewing hypothesis tests, but it has some theoretical problems that should be recognized.

One- and Two-Sample Tests

This unit on subchronic toxicity studies has dealt with simultaneous comparisons among all treatment groups. For reasons to be discussed below, this author does not believe that one- and two-sample comparisons are appropriate in this setting; however, they are widely used in the toxicological literature, so some discussion of them is called for.

One-Sample Tests

Suppose we wish to determine whether there have been changes in some measurement over time for a particular group of animals where the changes are more than random noise and represent some underlying change in the mean. We can set up the null hypothesis that the mean change is zero and construct a test based upon the paired differences for individual animals. If, for instance, we have 50 animals and all of them show an increase in SGOT level from week 12 to week 16, it is highly improbable that this is random noise. Under the null hypothesis of no effect, we would expect the number of increases to balance the number of decreases. What do we mean by "balance"? If 30 animals showed increases and 20 showed decreases, this is equivalent to having tossed a fair coin 50 times and obtained 30 heads. The probability of such a sequence is about 20%, so such a finding would not be improbable. If, on the other hand, there had been 33 increases and only 17 decreases, the probability of such a sequence is only about 3% and we could declare this as "significant" evidence against the null hypothesis. This type of test, where we simply count the number of increases versus decreases, is called a *sign test* (Hollander and Wolfe, 1973).

Slightly more powerful than the sign test in many situations is one that considers the actual increases and decreases. For instance, suppose there are 30 animals with increasing SGOT levels with changes ranging from 10 to 100 units and 20 animals with decreases with changes all less than 5 units. This case would be highly improbable under the null hypothesis. We can rank the differences by their absolute values and look at the average rank for the positive values (increases) versus that for the negative values (decreases). The null hypothesis induces a probability distribution for these average ranks in the same way we can compute the probability distribution for the Kruskal--Wallis test. This procedure is known as a *Wilcoxon signed rank test* (Hollander and Wolfe, 1973).

If we knew that the differences came from a Gaussian distribution but we did not know its variance, we could let the null hypothesis be that the mean of this distribution is zero. Then we could look at the ratio

Average difference/square root of the estimate of the variance

A modification of this ratio is known as *Student's t test* (Dixon and Massey, 1969) and the probability distribution of Student's t can be computed. Computer simulations and some theoretical work indicate that Student's t test, when applied to a set of paired differences such as this, is robust against the underlying assumption of a Gaussian distribution. That is, when one is working with paired differences and the null hypothesis that there has been no change, the p values computed for Student's t test are correct, regardless of the true underlying distribution.

Two-Sample Tests

Suppose we wish to compare 50 animals in the high-dose group to 50 controls with respect to changes in weight from week 12 to week 16. We could list all 100 values in rank order and compute the average rank for the high-dose group or that for the controls. The distribution of this average rank can be computed under the null hypothesis that all 100 numbers came from the same distribution. This is the *Wilcoxon rank sum test* (Hollander and Wolfe, 1973). This is very similar to the Kruskal—Wallis test. In fact, Kruskal and Wallis derived their test after Wilcoxon's work, in an attempt to generalize rank order tests from two to more than two samples. Another version of the rank order test for two samples was published independently of Wilcoxon by Mann and Whitney, so the Wilcoxon test is sometimes referred to as the *Mann—Whitney test*.

If we assume that both sets of data came from Gaussian distributions with the same variance but possibly differing means, we can estimate the common variance, usually called the *pooled variance estimator*, and consider the ratio

Difference in averages/square root of the pooled variance estimator or a slightly different version called the *two-sample Student's t test*. In his initial derivation, Gossett (who published under the pseudonym Student) considered only the one-sample situation, but this was an apparently obvious extension, and it has become widely used as the standard method of comparing two sets of data. However, there are some serious theoretical problems that have been only partially resolved which arise when dealing with the case where the two sets of data have differing variances or where one or two of them are not Gaussian. If, for instance, it is possible for one set of data to have an unboundedly small variance, a Bahadur—Savage-type distribution lurks in the background. Furthermore, computer simulations suggest that the two-sample t test may not be robust for asymmetrical distributions.

The two-sample t-test is really a version of Fisher's ANOVA for two groups, just as the Wilcoxon rank sum test and the Kruskal—Wallis ANOVA are related. So the theoretical problems of the two-sample t-test are also those of analysis of variance. However, in a situation where more than two groups are involved, use of the two-sample t-test to investigate differences between specific groups carries other dangers. Each comparison between two groups, by a two-sample test, has an alpha level at which significance is declared. However, if the study involves 1 control and 4 doses of test compound, then there are 10 possible comparisons of pairs of groups. Each one running at a 5% alpha level means that the probability of at least one type I error can be as high as 50%.

On the other hand, when the set of differences among the five treatment groups is examined with an analysis of variance (Fisherian or Kruskal—Wallis), the overall probability of a type I error is fixed at alpha. All comparisons between treatments that are made as contrasts within that overall test can be run without increasing the probability of a false positive (type I error).

When we run an overall analysis of variance, we are really allowing for all possible contrasts to be run within the alpha level. If we decide in advance that only a small set of contrasts is of interest, then we are in a situation of restricting the class of alternatives and we can increase the power by using other tests. For comparisons between controls and individual doses (but not comparisons between doses), there is a modification of the t test due to C. Dunnett called *Dunnett's t* (8). If the only contrast that will be run is one testing for a dose response, then the analysis of variance can be modified to produce a t test which will make that one comparison.

Analysis of Covariance

Fisher's Gaussian analysis of variance can be put into a more general structure than in the earlier section on Fisher's ANOVA. This will enable us to examine more complicated situations. To do this, we have to think in terms of *mathematical models*. For the purposes of this book, a mathematical model is an equation or a set of equations that describe the relationships between different variables. As an example, consider the situation where we wish to examine the relationship between liver weight, total animal weight, and treatment. If we designate the observed liver weight by Y, the observed total animal weight by X, and the effect of the ith treatment by the argument i, we can write the following set of equations:

Controls $Y = a(1) + bX$

Low dose $Y = a(2) + bX$

Middle dose $Y = a(3) + bX$

High dose $Y = a(4) + bX$

Because of random noise, not all of the control animals will have liver weights that fit the first equation exactly, and the scatter of values around that first equation can be used to estimate the variance associated with the random noise for the controls. Similarily, we can compute a variance estimate for the random noise of the low-dose liver weights about the second equation, and so on. We do not know the true values of a(1), a(2), a(3), and a(4) or of b, but we can estimate them by using techniques developed by Fisher.

Under the null hypothesis of no effect, a(1), a(2), a(3), and a(4) are all equal, so the set of four equations reduces to a single equation,

All groups $Y = a + bX$

Again, we can estimate a and b and estimate the variance of the random noise about this model. If there is a true difference between the groups, then that difference will be forced into this second estimate of the variance of random noise, and this estimate of the variance will be increased. If we call the second version model 0 and the original set of four equations model 1, then we can test the null hypothesis that model 0 is true by examining the ratio

Estimated variance of model 0/estimated variance of model 1

Fisher showed that this ratio has the same probability distribution as that of the F test in his Gaussian analysis of variance. In fact, if we had written the equations without reference to the total animal weight X, we would have had the mathematical model for analysis of variance.

In another direction, we could have made the model more complicated. We might have proposed different values of the parameter b for each treatment group or a quadratic or cubic function of X instead of a linear one. In general, we can think of a nested set of models. The outermost model is the one that is the most complicated, with the greatest number of free parameters. As we restrict these parameters to certain values (such as requiring that all treatments have the same value of b), we produce ever-simpler models nested within each other. We can make any one of these nested models our null hypothesis and test that null hypothesis against an alternative that consists of one of the more complicated outer models. The estimated variance off the null hypothesis model is divided by the estimated variance off the outer model to produce an F test of the null hypothesis.

This general procedure is used to test complicated models involving the introduction of possible effects due to the placement of cages in the holding room, varying times on study, differences between studies, and so forth. A number of elaborate models have become standard in such areas as agricultural research, and some of these have been adapted to toxicology. These techniques go by various names, depending upon how they originally arose: Sometimes they are called *regression* and at other times *two-way analysis of variance* or (more general) *multiway analysis of variance* or *analysis of covariance* (Dixon and Massey, 1969). When most statistical calculations were done on desk calculators, the distinctions between these methods and names existed because of the different algorithms that were used in the actual calculations; however, with the advent of computers the overall theoretical unity of these methods has led to the use of what is called the *general linear model*. And so, all such techniques are subsumed in large general-purpose computer programs that can be found in statistical packages such as BMD-P, SAS, SPSS, GLIM, and MINITAB.

But, as indicated earlier in this chapter, there is no "free lunch" in statistics. What assumptions are involved and what is lost in this general procedure? One of the assumptions is that the random scatter of values around the true model has a Gaussian distribution with constant variance from one treatment to the next. The other assumption is that the set of nested models contains an adequate description of the "truth" at some point. If, for instance, the animals consist of those that tend to develop a lesion and those that are resistant to that lesion and if the effects of treatment on these two types of animals are different, the nested set of models may make no adequate provision for such a situation unless the animals can be identified and factors for their response included.

Since these nested models can get very complicated, the user is often led into accepting the results without a critical examination of the assumptions. The best method for checking the validity of the models is to fit the observed data to the various models, examine how well the data actually fit, and try to interpret all the estimated parameter values in terms of the scientific problem at hand.

Improper Uses of Hypothesis Tests

In spite of their widespread use, hypothesis tests are really not the most appropriate statistical tool for most problems that arise in toxicology. Most studies in this field assume that the tested compound will be toxic at some dose and the study is mounted to determine the nature of the induced toxicity and the shape of the toxic dose response. If we are to approach these questions directly (as is done in estimating the LD_{50}), then we should be using statistical estimation procedures and coming up with confidence bounds on quantifiable components of the problem.

Unfortunately, hypothesis tests are often used in the toxicological literature to determine "safe doses" by comparing the effects of each dose to controls and declaring the highest dose at which there is no significant effect the "no-effect" dose. This involves the use of hypothesis tests to accept the null hypothesis and, as shown above, whether a statistical test will

come up significant is a function of the true underlying effect,
the number of animals used, and the statistical decision rule.

In a highly simplified scenerio, let us assume that the compound induces some liver toxicity that can be measured by
changes in SGOT level. Since there are always slight rises in
liver enzyme levels due to the metabolism of xenobiotic material,
we can expect to find slight rises in SGOT associated with any
dose of the compound. Thousands of animals may be required
for detection of the effect at a very low dose, but we can usually assume that it is there. In this case we can think of a
response of SGOT to dose as shown in Figure 3. At some
point the rise in SGOT begins to reflect liver damage and the
toxicologist should have a good idea of what that level might be.

In this scenerio the problem of finding a "no-effect" dose
becomes one of determining where the dose response crosses the
toxic effect line. As with estimation of the LD_{50}, we can con-

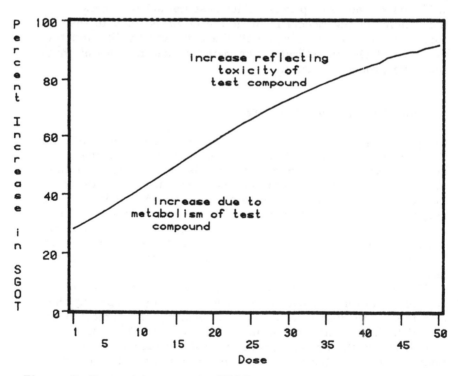

Figure 3   Percent increase in SGOT versus dose.

struct confidence bounds on the dose response curve and find
bounds on the "no-effect" dose as in Figure 4.

It will take more complicated mathematics to address the
question of dose responses for specific lesions, but the general
principle holds. If we can quantify the degree of effect and
include effects that are subtoxic, then we can construct dose
response curves and bounds that provide estimates of "no toxic
effect" doses that make both biological and statistical sense.

Another improper use of hypothesis testing often found in
the literature is the attempt to trade off one hypothesis test
against another. One finds, for instance, statements that the
test for dose response was significant but a two-sample comparison between controls and high dose was not, so the significant
dose response must be a false positive. As was indicated in the
sections on Neyman-Pearson hypothesis tests, there is an elaborate statistical theory dealing with the power of hypothesis tests.

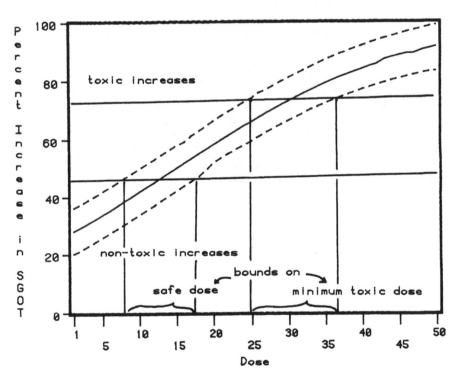

Figure 4 Percent increase in SGOT with confidence bounds.

A dose response test is more powerful than a simple two-way comparison, and so one would expect to find one showing significance more often than the other.

Furthermore, it can be easily shown that some types of tests are inherently more powerful than others. A Wilcoxon test is more powerful than a sign test. If we wish to be sufficiently dense, we can always find a statistical test that is so lacking in power that it has little chance of finding significance. It would make no sense to propose such a test as "contradicting" a uniformly most powerful test.

A final improper use of hypothesis tests occurs when they are used to compare a large number of events in a study. If the compound has a strong toxic effect, then we might expect that many of the measurements taken from living and dead animals will show something, and so an accumulation of statistical significances is sometimes used to increase the strength of the evidence. But if the test compound has no effect at all, we can expect the large number of tests to produce at least one significance with a high level of probability. If you deal enough poker hands from a deck of cards, reshuffling after each one, you will eventually deal a straight flush.

If the toxic effect will be manifested in a single organ, then statistical tests run on many organ systems will not improve the power of the test, although they will increase the probability of a false positive. Thus it behooves us to use a small number of most powerful tests, each of which concentrates on a biologically reasonable path of toxicity. Or, this author believes, we are better off abandoning the use of hypothesis tests entirely and concentrating on developing continuous measures of toxicity which can be used for estimation.

UNIT III
Chronic Toxicity Studies

Unit II

Chronic Toxicity Studies

7

Chronic Toxicity Studies

Overview

In terms of the general statistical problems posed, there is not much difference between chronic toxicity studies and subchronic ones. Both studies accumulate measured data such as weekly weights on living animals and both count the numbers of lesions seen on necropsy. However, in the subchronic studies there is seldom the need for statistical analysis of pathological lesions. For these studies the animals are usually killed while in robust health and any lesions can be clearly related to treatment, based upon knowledge of the expected biological activity of the test compound and its metabolites and observations made on the living animal.

It would pay to review how the toxicologist determines a "no-effect" or a "minimum toxic effect" dose from subchronic studies used in drug research. The highest dose is usually chosen to induce toxic damage, and the high-dose animals are examined for specific lesions that might be attributed to the compound at test. This is done by looking for lesions not usually seen in controls and that might be expected from the pharmacology of the compound. Once these lesions are determined, the lower-dosed animals are examined for indications of these lesions or putative precursors. For instance, if the compound induces stomach ulcers at a high dose, the toxicologist will look for inflammation at the lower doses.

A no-effect dose is then one at which neither the lesions associated with the compound nor any of their precursors are seen. A minimum toxic dose is one at which the first clear-cut incidence of the precursors, along with some possible lesions, is seen.

The most widely used statistical techniques have only a peripheral role in this type of decision making. This is probably because the toxic manifestations of a test compound in a subchronic study usually carry very little uncertainty.

With chronic studies, all animals, including controls, show some pathology at final sacrifice, primarily because they are much older. The problem of locating lesions associated with treatment becomes more difficult, and statistical techniques are often called upon. Unfortunately, the most widely used methods do not mimic the traditional toxicological examination of subchronic studies; instead, a standard statistical model has often been used that may not be appropriate to the problems at hand.

The following two sections will describe the standard statistical model and methods that have been developed to examine the incidence of specific lesions at necropsy. Subsequent sections of this chapter will examine possible alternatives.

Comparing the Incidence of a Single Well-Defined Lesion

The statistical model usually used in the analysis of chronic toxicity studies assumes that attention is focused on a single well-defined lesion, such as a palpable mammary adenoma. It ignores the severity of the lesion and whether it occurs only once or more than once in a given animal. Instead, it assumes that each animal supplies a single bit of information, whether it has a lesion or not. Thus the information used in the statistical analysis consists only of the numbers of animals at risk in a given group and those found to have the lesion in question.

We assume that there is a constant probability of lesion occurrence across all animals at risk in a given group. We also assume that the occurrence of the lesion in one animal has no influence on whether that lesion will occur in any other animal in the group. These are the conditions necessary for the random pattern of events to follow a *binomial distribution*.

Suppose, for instance, that we had 50 animals in each of four groups, with the following incidences:

Treatment	Number of animals with lesion(s)
Controls	12/50
Low dose	16/50
Middle dose	28/50
High dose	24/50

Since we know how many animals were at risk, only the underlying probabilities of lesion have to be estimated from the data. If there were no difference between the groups, there would be a single underlying probability, which would be estimated by taking the total number with lesions divided by the total at risk,

80/200 = .40

If the true probability of lesion were .40, we would expect to see around 20 animals with the lesion in each group; owing to random noise, this might vary between 13 and 27. Thus, although a pattern like the one above is suggestive of an effect, it might be due to chance. To test whether this is within acceptable chance variation, we can run a *chi-squared contingency table test* (Dixon and Massey, 1969). This would result, in the situation above, in a chi-squared statistic of 8.33 and a p value of .04. Thus, if we choose 5% as a cutoff to declare "significance," this pattern provides significant evidence against the hypothesis of no treatment-related effect.

As in the analyses of variance used in Chapter 4, it is possible to construct a directed test if we are allowed to assume that any reasonable alternative will have a dose response. The most general test for dose response is an adaptation of Bartholomew's test referred to in Chapter 4; however, there is a simpler test that is much easier to put on a computer and which is used more often: the *Armitage—Cochran test for linear trend in proportions* (Armitage, 1955). It is similar to the contrast for linear trend used in analysis of variance. This author's experience has been that the Armitage—Cochran test almost always produces the same significance levels as the Bartholomew test.

Because it is a directed test, the Armitage—Cochran test is more sensitive to data that implies a dose response, so the p value for the example given above is less than .01. Note that the observed data show an apparent dose reversal, 28 out of 50 at the middle dose and 24 out of 50 at the high dose. However, this is well within what might be expected from random noise alone, even if there were a true increase in the probability of lesion occurrence from the middle dose to the high dose.

These tests, based upon binomial assumptions, can be used to weigh the evidence that there is a treatment-related effect; however, they cannot be used to identify a no-effect dose once it has decided that there is such an effect. This is primarily because of the small amount of information supplied by each animal. For instance, even if there were no animals with a given lesion at the low dose, the fact that we used only 50 animals suggests that the underlying probability of lesion could be as high as 7% and still produce 0 out of 50 occurrences with reasonable probability. This residual uncertainty holds when the lesion does not occur in the controls. When the controls have the lesion but there is a dose-related increase, a similar type of reasoning can be used to show that, unless it is a drastic drop, even a lower incidence at the low dose cannot be used to prove no effect.

A single test of dose response is probably the most appropriate statistical technique to be used in comparing the incidence of a single well-defined lesion; however, it has become the practice in much of the toxicological literature to also apply tests that compare one group of animals to another. The tests (all of which produce very similar results) most commonly used are the *Fisher—Irwin exact test* and the *corrected chi-squared test for a 2×2 contingency table* (Dixon and Massey, 1969). Because they use less than the full number of animals in the study and they are not directed tests, they tend to be less sensitive than an Armitage—Cochran test. Therefore there is little purpose in applying them in addition to the more powerful procedure.

One widespread use of these tests comparing two groups at a time has been to identify a no-effect dose by looking for the first "significant" gap between successive doses or the first treatment that is "significantly" greater than the controls. This is a misapplication of statistics, since a statistical test of hypothesis (and one of low power, at that) is used to accept the null hypothesis of no difference. As indicated above, tests based upon the binomial distribution are inherently incapable of identifying no-effect doses.

Another misapplication of these two-group comparisons that has occurred in the toxicological literature is to cast doubt on the validity of a significant dose response test by pointing out that none of the tests comparing specific doses to controls were "significant." Since these two-group tests are of lower power, this is not an unexpected result, even when there is a treatment effect.

When a statistical test of hypothesis is significant, what is signified may not have anything to do with the toxicity of the test compound. It could reflect some other biological activity or it could be one of the 5% of the times the result was purely a result of random noise. So it makes sense to examine the biological interpretation one puts on a "significant" result, but it makes no sense to go back with other statistical but biologically blind tools to question the first one.

Adjusting Incidence for Time on Study

Even a cursory examination of the results of a chronic toxicity study will show that it seldom happens that a single well-defined lesion is seen at necropsy and all animals have lived to the final sacrifice. So it becomes necessary to adjust the procedures described previously for events that are not part of the simple binomial model. One type of adjustment that has been used in the toxicological literature is to consider time on study.

Some animals die or are moribund and thus killed before the completion of the study and many of the lesions analyzed occur only at advanced ages. This introduces the possibility of bias in the analysis if we just count the number of animals with lesions versus the number initially in the study. This is a problem that has also occurred in epidemiology, and there are several solutions that have been used in that field. One solution is to view time as a covariate and adjust the model to include as a factor the number of weeks or days from the start of the study until the animal died. Unfortunately there is no general consensus in the statistical community of what is the "best" covariate model. One model that is widely used and which can be found in most large statistical computer packages

is the *Cox regression* (Cox, 1972, 1975). This model makes
some strong assumptions about the nature of the treatment effect, which will be discussed in Chapter 9.

Other models are based upon *life table* procedures. Life
table procedures were developed to obtain consistent estimates
of incidence probabilities when the data consist of large numbers
of people examined at different ages. The epidemiologist con-
structs something like Table 1, where each of the columns refers to a single age group (such as those between 15 and 19
years of age) and each row refers to a particular medically important event. Since the sample of individuals examined is not
fully representative of the entire population, the incidence levels
in each of the cells of the table are multiplied by the proportion
of the population that actually occurs in that age group. The
result is an "adjusted" incidence figure that modifies the observed incidence to match the true population proportions.

Actually, one can run a statistical test of hypothesis with
this technique. The population figures can be used to predict
the expected frequencies in each of the cells, and these frequencies can then be compared to those observed. For that
matter, any projected pattern can be compared to the observed
pattern as a test of that projection.

This is the technique used most often in adjusting incidence
figures from chronic toxicity studies. The algorithm in most
computer programs designed for this type of analysis is due to
Robert Tarone of the National Cancer Institute. In the *Tarone
adjustment* (Tarone and Gart, 1980) the time from the start of
the study to final sacrifice is divided into small intervals during
each of which at least one animal with a lesion has died. Under
the null hypothesis that there was no treatment effect we would
expect the probability of lesion occurrence to be the same in
all groups. So the probability of lesion occurrence is estimated
by considering the total number of animals still alive in all
groups during that short interval of time. From this we can
compute an expected number of lesions in each group based
upon the number of animals still alive in a given group. The
difference between the expected and the observed is accumulated across all time intervals, along with a measure of the variance of that difference. The sum of those differences is
squared and divided by the appropriate variance.

Tarone's algorithm can be adjusted to become a directed test
against a linear dose response, where the dose response pattern can vary in slope from one time interval to another. The
resulting directed test is the time-adjusted equivalent to the

Table 1   Example of a Life Table from Epidemiology: Birth Rate per 1000 Unmarried Women

Year	Age of Mother (years)					Adjusted Rate
	15–19	20–24	25–29	30–34	35–39	
1980	28.2	39.7	30.5	17.4	7.8	28.4
1978	24.9	35.3	28.5	16.9	8.2	25.7
1976	23.7	31.7	26.8	17.5	9.0	24.3
1974	23.0	30.5	27.9	18.4	10.0	23.9
1972	22.8	33.5	30.8	22.6	12.0	24.8
1970	22.4	38.4	37.0	27.1	13.3	26.4

Armitage—Cochran test. There are a number of variants of the
Tarone algorithm (most of them due to Tarone, who has investigated the theoretical background of this procedure) which use
different methods of computing expectation and different types
of dose response. They all appear to have similar power and
produce similar results. When this method is used to compare
two groups, it reduces to a version of the *Mantel—Haenszel procedure* (Mantel, 1963) for combining data from a number of different studies.

One problem with applying this type of analysis occurs if the
lesion in question can cause the animal's death. Then two types
of events can occur: The animal can die with the lesion but
the cause of death is independent of that lesion or the animal
can die from the lesion. In particular, the animals that die
prior to the final sacrifice will tend to die because of the lesion,
while those that survive until final sacrifice will be alive in
spite of the lesion. All of these statistical procedures do not
really examine the probability of an animal having the lesion;
they examine the probability of detecting an animal with the
lesion. Thus, if the lesion is the cause of death, the probability of detecting the lesion will be different than if the lesion is
incidental to the death. For this reason some statisticians have
recommended that the data be divided into those animals which
die because of the lesion and those which die incidental to the
lesion, that separate time adjusted tests be run on each set of
animals, and that the final p values be combined.

The difference between the two approaches arises from
definition of *animals at risk*. If the animals that die during a
given interval of time die from causes independent of the lesion
being counted, then they can be thought of as a representative
sample of all animals alive at the beginning of the period.
Since they are representative of all the animals, the percentage
of them with the lesion can be used to estimate the percentage
of animals in the entire group with that lesion, so the denominator used in the calculation is the number of animals that died.
If, however, they die because of the lesion, then the total number of animals alive at the beginning of the period must be used
in the denominator, since death due to the lesion is a marker
of the lesion and the animals that have died are not representative of the entire set of animals.

It is not often that the pathologist can make a clear-cut
decision on whether a death was due to a particular lesion, so
this recommendation is not often taken in practice. Further-

more, if most of the animals live to final sacrifice, the slight differences in probability and the small numbers of animals involved mean that the final computed p values will not be affected. Thus this recommendation may involve a more or less moot point.

There is another problem with the blind use of time-adjusted analyses, however, that occurs if one treatment induces early deaths. The statistical procedure tests whether there is a dose-related increase in the percentage of animals with lesion, but there must be animals available for occurrence the lesion. If the lesion is one that occurs late in life and if all or most of the high-dose animals die before the lesion can occur, the statistical test will probably find no significance. This is because there was no dose response in the observed incidence, but the lack of significance is obviously the result of asking the wrong question.

It can be seen, then, that time-adjusted analyses are not an automatic panacea to the problems created by early deaths. For every statistical test run the toxicologist should be careful to examine the biological hypothesis tested.

Analyzing a Single-Graded Lesion

In practice, it seldom occurs that there is a well-defined lesion whose incidence can be compared across groups. In examining tumors, for instance, the pathologist is often hard put to determine whether the tumor is malignant or not. At one end of the spectrum tumors merge into hyperplastic nodules, and hyperplastic tissue merges into displasia. When the statistical analysis demands that the pathologist make an absolute determination of whether this lesion is there or not, the resulting p values may be influenced more by the pathologist's morning meal than by the sophisticated statistical adjustments that were run on the data.

In fact, it has been traditional practice in toxicology to grade many lesions. This is done in order to establish a pattern of precursor lesions when attempting to find a no-effect dose. It would seem sensible to include these gradations in the statistical analysis. The grading of lesions involves a certain amount of subjectivity. When two or more pathologists examine the same set of tissue specimens, they often disagree on which of two contiguous grading scores should be used, but they seldom disagree by more than one score. Thus the grading of a lesion can be thought of as a measurement of effect which contains a small bit of random noise due to differences between pathologists' opinions.

If we were to use all the skill of the pathologist, we should encorporate these grades of effect into the statistical analysis and have each animal supply more information than a simple occurrence--nonoccurrence decision; instead, each animal should supply a grading score. In clinical research such subjective grading scores are often used to allow the attending physician to evaluate the usefulness of the experimental treatment. They are usually analyzed in one of two ways. Either the numerical value of the score is taken and techniques applied for continuous measures, such as the analysis of variance, or the number of patients in each grade of response is counted and contingency table tests applied to the resulting pattern of counts.

The same thing can be done with the severity scores of the pathologist in toxicology. Both techniques described in the previous paragraph tend to produce very similar answers, so let us consider using the score as a numerical measure of severity. In some cases, such as those involving tumors, we can line up a sequence of putative precursor lesions for scoring. As an example, tumors and their precursors can be scored so that 0 stands for completely normal tissue, 1 for slight dysplasia, 2 for moderate dysplasia, 3 for hyperplasia, 4 for hyperplastic nodules, 5 for encapsulated benign tumors, 6 for benign tumors of a more life-threatening nature, and so on.

Once a lesion and its precursors are scored, the data can be subjected to analysis of variance, and confidence bounds can be computed for contrasts between controls and each dose. The toxicologist is now free to define a no-effect dose as one whose upper 95% confidence bounds is less than some predetermined value.

All of the above is somewhat speculative. The author has tried this sort of analysis only a few times, and the results has not always been in agreement with the opinion of experienced toxicologists. It is a statistical method designed to mimic a procedure of biological evaluation that is done very effectively now for subchronic studies. Before it can be adopted as a useful tool, it should be tried out in a larger number of cases and skeptically examined.

Comparing Multiple Lesions

One problem with the data from a chronic toxicity study is that most of the animals die or are sacrificed at old age. As such, even for controls, their pathology usually includes a large number of senile lesions. Even subclinical diseases associated with old age and death in any species are usually complicated syndromes, involving many organs and inducing a complex pattern of pathology. Thus a diabetic human being may die with many cardiovascular lesions plus peripheral nerve damage. Typically, an aged rodent will suffer from a pattern of kidney disease with related pathology in the urinary tract.

Because of this tendency to complex syndromes of disease, it is simplistic to examine only one lesion at a time in a chronic toxicity study. As the test animals age, different syndromes emerge. If the treatment has biological activity, then it can be expected that this activity will influence the frequency of specific syndromes and the patterns within individual syndromes. Thus the most comprehensive analysis of the pathological findings should be directed toward measuring degrees of derangement among normally occurring syndromes.

The best way to do this is to establish what those syndromes are as part of the science of toxicology. Statistical techniques would then be applied to these biologically sensible combinations of lesions. Unfortunately this has not yet been done. It has been frequently noted, for instance, that there is a negative correlation between the occurrence of malignant lymphoma and hepatic tumors in mice. This suggests that these two signs are prominent markers of mutually exclusive syndromes. In the examination of a large study involving sugar alcohols (Salsburg, 1980), this author found statistically significant patterns that suggested mutually exclusive syndromes marked by adrenal lesions and subcutaneous tumors.

Outside the range of statistics, we need to establish what the more frequent syndromes of old age disease are among the animals usually tested. This can be done by either considering the biological deterioration of the species, statistical clustering techniques, or a combination of both. All of this remains to be done, and so this possible use of statistical techniques to more firmly define the nature of chronic toxicity remains no more than a wish.

If we could identify such patterns, what would be appropriate statistical methods of analysis? As indicated above, two ma-

jor questions must be addressed: What is the nature of the
chronic toxicity induced by the treatment and can we define a
dose response or at least locate no-effect and minimum toxic effect doses?

Let us assume that we have identified a pattern of lesions
that can be attributed to treatment. Then we could decide
whether a given animal has that pattern or not and go back to
counting animals with specific patterns versus animals at risk.
The methods of statistical analysis would be the same as those
used for counting animals with specific lesions. However, this
would not make use of the biological insights gained by identifying mutually exclusive syndromes of disease. Since every animal must eventually die, we can think of the syndromes of disease as exhausting all possible patterns (even if we have to include a "syndrome" made up of "everything else"). Since statistical procedures are so widely used in other fields, it turns
out that there is a well-developed statistical procedure that can
be adopted to this situation: the method of analyzing mixtures.

The theory of mixtures was developed to deal with problems
in chemical engineering. In a mixture the material is organized
in terms of relative proportions of different components. For
instance, if we increase the proportion of iron in an alloy, we
have to decrease the proportion of something else. In the
same way, the "normal" pattern of senility in an animal might
consist of a mixture of syndromes, with different probabilities
of one syndrome over another. If the treatment has an effect,
we can expect it to change the probabilities (or the proportions) of one or more syndromes. Using the techniques already developed in engineering statistics, we can test for
changes due to treatment, estimate dose response patterns, and
get confidence bounds on differences (in order to identify no-
effect doses).

If it is possible to identify degrees of pathology associated
with specific syndromes, then these can be added as measures
of severity and the data subjected to standard statistical
methods of multivariate analysis or we can further subdivide
the collection of exhaustive syndromes into subsyndromes of different degrees of severity and continue to use the tools of mixture statistics.

The major point here is that the problems of dealing with the
entire animal (rather than counts of arbitrarily defined lesions)
lie more in the realm of biology than in statistics. Once the
toxicological community can agree on classes of syndromes, the
statistical techniques are at hand, waiting to be called into use.

Use of Historical Controls

All of the techniques of statistical analysis discussed in this chapter remain ignorant of history. They view the data on hand from a single study as if they were the only data available and judge significance on the basis of purely internal evidence. Most toxicologists have access to large bodies of previous studies, all of them with controls. It seems foolish not to use these historical controls in the analysis of the current data.

There are *Bayesian* techniques in the statistical literature that appear to be designed for just such a situation, and several articles have appeared in statistical journals suggesting Bayesian procedures that can incorporate historical controls in chronic toxicity studies. All of these papers have assumed that there is a single well-defined lesion being examined, and so they represent attempts to modify the simplistic binomial model.

In the binomial model the only unknown parameter is p, the underlying probability of lesion occurrence. The current controls give us some information about that probability, but with only 50–100 animals this information will contain a considerable amount of uncertainty. The statistical tests that compare controls to treated animals have to take this uncertainty into account and are less powerful than they would have been if we knew for sure the p value for the control group.

The Bayesian approach is to consider the probability p of lesion occurrence for the control group as a random variable that varies from study to study. In this case the estimated p values from previous control groups provide us with some idea of the probability distribution of the random p. The observed data from the current study are then used to improve upon the knowledge of that probability. The result is an estimate of the probability of lesion in the controls that includes both sets of information. Theoretically, the treated groups can be compared to this modified estimate to test the hypothesis of no treatment effect.

Bayesian techniques were developed in the statistical literature primarily for purposes of estimating parameters (or, rather, the probability distribution of parameters) and applications of these techniques to hypothesis testing have not been firmly defined. So the applications of Bayesian techniques that have appeared so far in the literature produce different results when applied to the same data, depending upon how the initial distribution is described and how the final distribution is converted into a test of hypothesis.

If we accept the Bayesian approach that the underlying
probability of lesion occurrence is random and changes from
study to study, then one of the uses to which toxicologists
would like to put historical controls is lost. Suppose a study
shows a significant dose-related increase in some lesion from the
controls to the treated groups but the latter have the same incidence of lesions as was often seen in historical controls and
the current controls have a much lower incidence. One is
tempted to reject the current controls as being "too low" and
note that the treatment induced no more lesions than is normal.
However, the Bayesian approach says that p for the controls is
a random variable, so it is conceivable that for this particular
study there was an unusually low probability of lesion in untreated animals. However, the same unusually low p should
have held for the treated groups, if there were no treatment
effect, so the significant finding is valid.

An alternative use of historical controls is to compare current controls to what has been seen in the past. If there are
drastic differences, this suggests that this study was different
from the others in some way. Statistical techniques can be
used to determine if this difference is more than might be expected by chance, given the previous patterns of results.
But, once it has been decided that the current controls are
"significantly" different from past controls, it becomes the job
of the scientist who ran the study, the toxicologist, to decide
how that difference can be used in interpreting the results.

As an added complication, the variability of historical controls may involve trends. Some of these are long-term trends
that reflect genetic drift or changes in animal husbandry practices; others may be seasonal trends. Any attempt to use historical controls must include an analysis of the historical data
for such trends. If trends occur, the safest approach is to
ignore the historical data as unreliable for judging current
events. It is theoretically possible to estimate those trends and
use them to predict the current control p; however, such techniques will tend to be very dependent upon the arbitrary
mathematical models used in the analysis. As Paul Levy, the
probabilist, once said, "Prediction is very difficult, especially
the future."

8

Lessons from Chronic Toxicity Studies

Use of Count Data Only

Although a full necropsy on an animal includes a great deal of information that can be used in determining the toxic profile of a treatment, most statistical procedures now in use reduce that information to a single yes or no answer for each animal, and the statistical analysis deals only with numbers of animals with specific lesions. This destruction of information reduces the ability of statistical analysis to answer the more important questions of toxicology.

When each animal is reduced to a simple yes or no, studies with 50 or 100 animals in a group are not very powerful. It is impossible for statistical techniques to distinguish effects of treatment if there is a low incidence of rare lesions, since it takes a minimum of six animals with lesions, all of them in the high-dose group, to produce even 5% significance. Although the lesion may be so rare as to indicate an obvious treatment difference, that knowledge is not put into the statistical analysis and must be applied by the toxicologist in spite of a statistical conclusion of "no significance."

The lack of power shows up even if there are enough animals with lesions to make it possible to find significance. With only 50–100 animals per group, the increase in the number of animals with lesions from controls to the high-dose group has to be almost 50% before we are reasonably sure of being able to note significance.

Thus, the first lession is the following:

1. Reducing the information on each animal to the simple occurrence or nonoccurrence of a specific lesion makes it impossible to detect treatment-induced tumors of a rare type and difficult to detect slight increases in other lesions that might be due to treatment.

The use of such minimal information from each animal also makes it necessary to use simple mathematical models that have no provision for sophisticated questions. The simple binomial model usually used has only one unknown parameter, the underlying probability of lesion occurrence in a given group. Because low but nonzero probabilities can lead to counts of zero animals and because comparing counts in two small groups of animals cannot distinguish between underlying probabilities that differ by 2–3%, there is no way of using these procedures to determine no-effect doses. They can be used only as hypothesis tests to detect treatment-related effects.

From this we have the second lesson:

2. Using counts of animals with lesions as the only data, it is impossible to determine no-effect or minimum toxic effect doses.

Asking the Right Questions

Although the first uses of statistical methods in the analysis of chronic toxicity studies involved only counts of animals with particular lesions, it soon became apparent to the statisticians working in the area that early deaths during a trial will have an effect on the final conclusions. As a result, life table techniques, already used in epidemiology, were adopted to these studies, and analyses of effects were "adjusted" for time on study.

However, any statistical "adjustment" is nothing more than the imposition of an arbitrary mathematical model on the data, so that parameters of the model that are independent of the treatment can be estimated and their influence can be subtracted from the observations. When a mathematical model is imposed on biological data, it is important that the biological scientist (the toxicologist) understand what the parameters of that model mean in "real life." For instance, if the early deaths are a result of treatment and if the lesion examined is a late-occurring one, the adjustment for time on study will not be able to answer the question of whether the treatment induced an increase in the incidence of the lesion; instead, it answers another question that has little or no toxicological meaning.

Here the lesson to be learned is the following:

3. The nature of statistical "adjustments" has to be understood by the toxicologist and must make biological sense before it can be assumed to answer the question posed.

Asking the right questions also involves taking a careful look at the type of data actually collected and how that entire set of data can be used to answer the toxicological questions of what kind of effect the treatment has and at what doses. The toxicologist has to answer these questions in terms of the entire animal, and not just in terms of specific lesions. Statistical techniques might help, but only if the statistician knows about all the data.

Aspects of the data not normally used now in statistical analyses involve degrees of severity, multiplicity of lesions, and interrelationships between lesions (syndromes of senile disease). Since most lesions have different degrees of severity, it is probably a mistake to demand that the pathologist make a single yes or no statement about the existence of a lesion in a specific animal. In this case the pathologist's rule for decision making when in doubt may have more of an effect on the final statistical significance than any of the statistician's mathematical manipulations. However, there are standard statistical techniques for handling measures of lesion severity, and these can be applied so as to reduce the influence of borderline decisions.

Treating each animal as a simple yes or no also destroys the information that some animals have more than one lesion, either of the same type or of different types. As a result, the toxicologist is faced with a statistical analysis that provides significance levels for every lesion examined. He or she must decide what this signifies for toxicology. If all of the significance tests were statistically independent, then 1 out of 20 will be "significant" by chance alone, and running significance tests on 10–20 lesions will greatly increase the chance of a statistical false positive.

Statistical false positives often look like real events. By chance alone the incidence of lesions lines up in a highly suggestive dose response pattern and there might appear to be a considerable difference between the controls and the high-dose group. So there is often no way of knowing from the data if the result was a spurious significance or due to a real effect. For studies that take less time and effort, it is possible to run a second or third study to verify a result; however, the time and great expense required by a chronic toxicity study mean that there is seldom an opportunity to replicate.

One way of reducing the probability of a statistical false positive is to correct the computed p values to meet the requirements of a *Bonferroni Bound*. If we run six statistical tests and set the level of significance at 5%, the probability of a false positive is no greater than 30% (6 × 5%). This bound holds whether the tests are statistically independent or not, since it is based upon a fundamental mathematical property of probability known as countable additivity. However, it is a bound, which means that the mathematical theorems only tell us that the probability has to be less than this. Another bound, equally valid, is to say that the probability of a false probability has to be less than or equal to 100%, regardless of the method used. This second bound is just as valid but less useful, because the Bonferroni bound is usually less than 100%.

One way to use the Bonferroni bound to avoid a high probability of a statistical false positive is to use a nominal "significance level" small enough so that the bound keeps the true significance to less than 5%. This can be done with a nominal "significance" of .05/n, where n is the number of tests run in a given study.

Driven to an extreme, use of the Bonferroni bound can make it virtually impossible to use statistical tests of hypothesis on a chronic toxicity study. All we need to do is routinely run enough tests so that only the most extreme and obvious patterns are deemed "significant."

Another way to make the statistical tests useless is to cut the lesions into small subsets based upon morphology and organ. For instance, we might run separate tests on each type of brain tumor, depending upon the tissue type involved, or we might subdivide leukemias or lymphomas by the organs involved, or adenomas by the tissue source in the fetal layer. All of these are very reasonable ways of examining the pattern of tumors in order to estimate the nature of the toxic effect, but all of them divide the data in such a way that the analysis of each lesion has a low probability of showing significance, and the large number of analyses require a larger correction based upon the Bonferroni Bound.

This leads to lesson four:

4. Until all the available data can be put into one overall statistical model for analysis, the interrelationships among components of the data may be interpreted by the statistical techniques in a manner that will make the statistical analysis worthless to the toxicologist.

Biology Versus Statistics

Statistical methods as now applied to chronic toxicity studies are not as effective in answering the toxicological questions posed as they might be. This is primarily because the mathematical models involved are not adequate to mirror the biology. This not for a lack of standard statistical techniques. There is a large body of statistical thought that has not yet been brought into the analyses of these studies but which has been shown to be quite effective in other fields.

Some of these techniques were touched upon in Chapter 7, and they include mixture techniques from chemical engineering, Bayesian techniques, multivariate procedures, and scaling devices from sociology and clinical research. This author believes that this is because the communication between the statistical community and toxicologists has not been as open and interactive as it should be. Thus we have the fifth lesson:

5. The usefulness of statistics in chronic toxicity trials is far from ideal, and the currently used statistical models need a healthy injection of biology.

9

Theory Behind Chronic Toxicity Studies

2 × K Contingency Tables

Assume we count the number of animals in each of four treatment groups that have malignant tumors. The resulting data can be arranged in a contingency table as follows:

	Control	Dose Low	Dose Middle	Dose High
With tumor	12	11	24	26
Without tumor	38	39	26	24

For each group of animals the number of them with malignant tumor is assumed to follow a binomial probability distribution with n = 50 but p unknown (see Chapter 3 for a detailed description of the binomial distribution). Analysis of this table deals entirely with the four unknown p values.

If we wish to determine whether this pattern is the result of a treatment-related effect or can be attributed entirely to chance, we set up the null hypothesis that all four groups have the same p. If the pattern observed is highly improbable under this hypothesis, then we reject it in favor of attributing some effect to treatment. The effect of treatment is assumed to be determined entirely by a change in the underlying p from controls to each dose.

Since we do not know the true p values, we are forced to estimate them from the data. Under the null hypothesis (one constant p), the "best" estimator (maximum likelihood) is the proportion of animals with tumor across all four groups,

73/200 = .365

If this were the true p for all four groups, we would expect to have 18 or 19 animals with tumor in each group. The exact expectation is 18.75. The standard chi-squared analysis of such data compares the observed frequencies to this expectation and adjusts for the variance associated with a binomial with n = 50 and p = .365. For this table the resulting chi-squared statistic has a computed p value of less than .005. Thus we can take this pattern as supplying significant evidence against the null hypothesis and there seems to be a relationship between the treatment and the probability of finding an animal with at least one malignant tumor.

But suppose the result had been as follows:

	Control	Dose		
		Low	Middle	High
With tumor	12	11	14	18
Without tumor	38	39	36	32

A chi-squared contingency table test yields a chi square whose p value is greater than .15, and we cannot use this data to reject the null hypothesis of no effect. However, the prior knowledge that any effect will be dose related suggests the use of a directed test.

The most general directed test is Bartholomew's (Barlow et al., 1972). It runs as follows. If there is a dose-related effect, then we would expect that the p values underlying the four binomials will not decrease as we go up in dose. To *not decrease* means that the p values can remain constant from one dose to another, so the original null hypothesis of no difference is part of the class of hypotheses addressed by Bartholomew's test. However, it does not include as a possible event a decrease in p from control to low dose or from one dose to the next.

The "best" estimators of these nondecreasing p values are computed by starting from the lowest dose (controls) and averaging with the next highest as long as the observed frequency in the next highest dose is lower and otherwise leaving it alone. This is done first with controls and the low dose, then controls, low dose, and middle dose, and then all four.

Under this estimator we would expect 11.5 animals with tumor in the controls and in the low-dose group, 14 in the middle-dose group, and 18 in the high-dose group—which is almost what we observed. The chi-squared statistic that compares the observed to the expected under this more general hypothesis is computed, and Bartholomew's test consists of the difference between the original chi-squared statistic for the null hypothesis and this newer, smaller chi square. Bartholomew worked out the probability distribution of this new test under the null hypothesis of no difference in p values. It involves a separate calculation for each choice of n and the number of treatments, and the calculations involve iterative procedures from advanced calculus. With modern computers this should not be a barrier to the use of Bartholomew's test. However, Bartholomew's test has not yet been included in standard statistical packages, so it should be looked up as a theoretical standard against which to judge the performance of other tests.

The test most widely used is the Armitage—Cochran test (Armitage, 1955). Instead of including all possible nondecreasing dose response patterns in the class of acceptable hypothesis, Armitage—Cochran includes only those whose p values can fit a linear relationship. Recall that a linear relationship requires that each p be of the form

$$p = A + BX$$

where A and B are fixed parameters of the distributions and X varies from dose to dose. The power of the Armitage—Cochran test depends upon the choice of X, the "measure" of each dose. If we use the actual dose in mg/kg per day, the test tends to depend more upon the original placing of dose and less on the results of the study. However, if we use coded doses of the form

0 for controls
1 for low dose
2 for middle dose
3 for high dose

then the Armitage—Cochran test tends to produce p values very close to those obtained from Bartholomew's test. In theory, the Armitage—Cochran test should be less powerful for dose responses that are not restricted to linear patterns, but this does not appear to be so. However, the Armitage—Cochran test is overly sensitive to linear patterns. A sequence such as 10/50, 12/50, 14/50, and 16/50 produces a p value that is very close to 5%. What it is tagging here as improbable is not necessarily a treatment effect but, rather, an unusual exact line-up of the numbers.

Both the Bartholomew and Armitage—Cochran tests are asymptotic tests; that is, the p values are computed for probability distributions that are approximations of the true probability distributions. These approximations get better as the number of animals at risk increases—hence the word *asymptotic*. However, if less than 10 animals have the lesion across all groups, the difference between the true p value and the approximate p value may be enough to make the difference between formal significance and nonsignificance.

It violates common sense to believe that a p value of .049 means a treatment "causes" the lesion while a p value of .055 means it does not; however, toxicology is often examined within the context of government regulations, and it is sometimes convenient to have clear-cut dividing lines. To aid in such bureaucratic decisions, it has been proposed that the Armitage—Cochran test be "corrected" so that the approximating p value will be closer to the true p value. The correction (due originally to Yates in agricultural research of the 1930s) subtracts a fraction between 0 and 1 from each observed count before the calculations are begun. Thus some computer programs will produce two p values, one uncorrected and the other corrected.

Similarily, with modern computers it is possible to compute the exact distribution of the Armitage—Cochran test statistic and obtain the "true" p value. Such a test is called an "exact" test. Since there is no reason to believe that the data follow the assumed binomial distributions to begin with, all this exactitude may be a case of getting precisely the wrong answer. It would make more sense to allow borderline significances to be a vague region of p values, where the biological judgment of the toxicologist has room to determine whether or not the purported effect is treatment related.

Adjusting for Time and Other Covariates

Two methods are used for adjusting for time on study: One is based upon an old statistical trick that involves finding a deviation from the expected and a variance for each of a large number of small subdivisions of the data and then combining these to produce a powerful test; the other involves modeling the effect of time (and other covariates) and allowing for that effect to be subtracted from the analysis before comparing treatments.

The first method forms the basis of the Tarone adjustments (Tarone and Gart, 1980). We divide the time course of the study into a large number of subintervals in each of which at least one animal died with the lesion in question. The best way is to organize the entire study across time and draw a line at each point where an animal died and was found to have the lesion. The number of animals alive in each group at the beginning of each period is the number of animals at risk during that period. Thus we obtain a 2 × K contingency table of the type examined in the previous section for each interval of time. Each interval yields the parts of an Armitage—Cochran test statistic. These parts are then combined into one overall test, which gives the number of animals expected to have the lesion under the null hypothesis of no effect, which is usually printed out in the computer programs that run this type of test.

None of the adjusted tests are really tests of whether the treatment induced an increase in tumorigenesis. The information we have is that an animal died or was killed moribund and found to have the tumor in question. Thus the Tarone adjustment provides a test of whether the pattern of animals that died with tumor differed from chance in the direction of a dose response.

We can think of tumorigenesis associated with early death as a continuum in the relationship between death and tumor. The

tumor might be immediately lethal, so all animals that die with
tumor are the only animals with tumor. Then the Tarone adjustment is a test of tumorigenesis, since death is equivalent to
tumor development. At the other extreme we can think of the
death as completely independent of the tumor. Then the animals that die are in some sense "representative" of all the animals that were alive at the beginning of the period. If so, then
the percentage of animals with tumor can be estimated by the
percentage of animals that died from causes unrelated to the
tumor. In this case the Tarone adjustment tells us nothing
about the incidence of tumor, only about the incidence of tumor
associated with death. If the treatment induces early deaths,
then we might find ourselves counting more animals with tumor
in the treated group and overestimating the percentage with
tumor.

It has been argued that the animals that die during the
course of the study should be divided into two groups, those
that die from the tumor and those that die incidental to the
tumor. With this division a Tarone-type analysis can be run on
the first group and an analysis using the number of animals
that died as the denominator can be run for the second group.
Another argument can be made that such a division is purely
arbitrary and that one gets a well-defined statistical test from
the Tarone adjustment, leaving it up to the toxicologist to interpret the results.

A second approach is to model the effects of time and other
covariates. The most widely used such model is the Cox regression model (Cox, 1972, 1975). We start with a concept
from reliability testing in engineering: the *hazard function*.
This is the probability that an animal alive on day t will die
and be found to have the lesion before the day is out. If the
lesion is an old-age tumor, then the hazard starts off low (few
young animals die with the tumor) and gradually increases until
(as with testicular tumors in Fischer rats) it may reach 100%.
On the other hand, leukemia in mice tends to have a decreasing
hazard function, that is, its incidence is high for young animals
and lower for older ones.

We can modify the hazard function as a probability by multiplying it by a function of other aspects of the study. For instance, if the treatment has changed the probability of lesion,
then this change can be modeled by introducing a parameter
for the treatment that is multiplied by the underlying hazard
function. We take on an assumption when we do this called the
factorability of the hazard function.

At this point we are dealing with a mathematical abstraction and it is difficult to find a biological interpretation for this assumption. In some sense we can think of the hazard function as being the result of a force inherent in the conditions of the study. If the treatment is another factor, then the treatment only increases the force at any point in time and does not modify the way in which that force changes over time. Suppose we are examining possible tumorigenesis. If we think of tumors as resulting from initial "hits" by exogenous tumorigens followed by a purely endogenous development, then a treatment that only increased the frequency of hits would be a factor in the hazard function.

One needs very large numbers of animals to be able to test the hypothesis that the hazard function is factorable. One such study—the FDA's ED-01 study—(ED-01 Task Force, 1981) provides evidence that for the tumorigen AAF and either liver or bladder tumors in mice the hazard function is not factorable.

However, in order to adjust the hazard function for time and other covariates, we need to assume something about its theoretical structure. If it is not factorable, then there are an uncountably infinite number of other possible models, each one as restrictive or more so than the assumption of factorability. So in the Cox model we assume factorability and hope that it will not induce too great an error in our calculations.

If we can introduce a factor for treatment, we can also introduce factors for other identifiable covariates, like initial weight or any time-dependent event, like a planned sacrifice pattern. The initial effect of time on study is taken up within the concept of a hazard function. The Cox regression methods let us estimate the initial (unmodified) hazard function in any fashion, including data-bound, model-free methods. The effects of covariates and treatment, however, are assumed to be factorable.

The result is both a test of the null hypothesis of no treatment effect and a means of estimating the time course of the development of the lesion and a dose response. Thus, if the intent of the study is to determine not only the effects of constant chronic exposure but also the effects of specific intervals of exposure, these techniques offer the chance of doing that.

Scaling Severity of Lesions

Suppose we can view a lesion or a sequence of related lesions as a continuous modification of a particular tissue. As an ex-

ample, we might go from normal bladder cells to displasia, to
mild hyperplasia, to obvious hyperplasia, to the borderline between hyperplasia and tumor, to benign tumor, to another borderline region, and to malignant invasive tumors. Or we might
consider degrees of severity of nephritis or gastric irritation
up to ulcer. Psychologists and sociologists often have to deal
with this type of data.

One of the problems with scaled data is that all we really
know is that the scale indicates an increasing involvement, but
we do not know the true "distance" between points on the scale.
We do not know if it is the same "distance" from normal tissue
to displastic cells as it is from severe hyperplasia to benign
tumor. If we assign arbitrary values of 0, 1, 2, 3, and so on,
for the components of the scale, we can deal with average values and percentages of animals with certain values and still be
able to interpret the results in terms of the original descriptions
of severity. However, if we start looking at differences over
time within a group or from one group to another, we cannot be
sure that a difference of 1.3 units means the same thing for
a treatment that averages 2 units as it does for a treatment that
averages 6 units.

It should be noted that there is nothing mathematically
"wrong" with treating the scale values as if they were continuous variables for each animal and using standard techniques
like Kruskal–Wallis ANOVA. Each animal is assumed to provide
a random variable taken from the set of possible scale values
and the statistical analysis deals with examining the probability
distributions of these random variables. The problem with scaling comes when we finish the appropriate statistical manipulations and attempt to interpret the conclusions in terms of the
original biological observations.

Fortunately, the toxicologist need not be as concerned with
the problems of scaling as the sociologist. Much sociological
scaling is done across different studies and observations taken
at different points in time, so the consistency of the scaling is
of great importance. If we restrict the analysis to a single
study, with the same pathologists scoring the slides, then the
scaling can be compared across treatments. Thus scaled values
can be used as if they were continuous measures to test the
null hypothesis of no effect against a restricted hypothesis of a
dose response.

If the statistical analysis suggests a treatment-related effect,
then the scaling can also be used to identify doses of no toxic
effect. This, however, will require biological decisions, and

the toxicologist will have to define some value of the scaling that
was not considered toxic. Then confidence intervals on the
mean scale for each treatment group can be used to find treatment doses whose confidence bounds are within the nontoxic
range. This assumes that some degree of the lesion is acceptable as nontoxic. If we are dealing with an exaggerated
pharmacological event for a drug (such as ulceration for an
anti-inflammatory drug), this is not much different from what
is done now, when the number of animals with toxic levels of
exaggeration is counted. If we are dealing with a supposedly
unacceptable event such as tumorigenesis, then it may not be
possible.

For an example of the use of scaling, Table 1 displays a set
of individual animal values taken from a study of the effects of
chronic xylitol ingestion. The male mice here that survived to
final sacrifice represent a group of controls and one dose of
test compound. For four organs, the lesions are scored as
follows:

0, no lesion
1, a lesion unrelated to tumor
2, a mild degree of hyperplasia
3, a greater degree of hyperplasia for bladder or a single
   adenoma for other organs
4, a benign tumor or tumors
5, a malignant tumor

These organs are the bladder, kidney, liver, and lymphatic
system. The final column scores for tumors or hyperplasia in
any other organ, taking the most severe occurrence.

This set of five numbers for each animal is called a *vector*,
and the degree of toxicity has to be inferred from the multiple
nature of this vector. For instance, among the controls animal
number 39 has a benign tumor in the bladder, malignant lymphoma, and a benign tumor elsewhere, with nontumorous lesions
in the kidney and liver. Animal 71, on the other hand, has
malignant lymphoma, a malignant hepatic tumor, a benign tumor
of the kidney, and hyperplasia elsewhere, but no lesion in the
bladder. We can think of each animal's vector as forming part
of a cloud of points in five-dimensional space. Through the
use of *multivariate statistics* we can determine if the cloud of
points based upon the control animals and that based upon a
given treated group differ by more than chance.

Table 1   Lesions Found in Male Mice at Terminal Sacrifice
Scaled 0-5 (no lesion to malignant tumor)

Part 1:   Controls

Score by site

Animal	Bladder	Kidney	Liver	Lymph	Other
1	0	0	0	0	3
3	0	0	0	0	4
4	0	0	4	0	1
6	0	0	4	5	0
7	0	1	3	0	0
9	0	1	1	2	1
29	0	2	4	0	4
31	0	2	2	0	1
32	1	0	1	2	2
39	4	1	1	5	4
62	0	0	4	5	1
63	1	1	4	0	4
67	0	2	4	0	4
70	0	0	1	5	4
71	0	4	5	5	1
76	0	0	4	2	5
78	3	0	4	0	4
79	3	4	4	5	0
85	0	0	1	5	4
87	0	1	4	5	2

Table 1 (continued)

Part 2:   20% Xylitol

| | Score by site | | | | |
Animal	Bladder	Kidney	Liver	Lymph	Other
408	0	1	4	0	1
412	3	0	0	0	1
418	0	1	1	0	1
422	1	1	4	0	0
423	0	0	1	0	1
424	3	2	1	0	4
431	3	0	1	0	4
432	3	4	4	0	1
433	3	0	1	0	1
436	3	1	4	1	1
438	2	1	4	1	4
439	2	0	1	0	4
446	5	0	1	0	4
449	2	0	4	0	4
445	4	0	4	0	4
461	3	0	4	5	4
466	2	1	1	0	0
474	3	1	1	1	0
477	2	2	4	5	1
478	0	2	4	0	1

Table 2 displays the average values of the individual components of the vectors for each of the treatment groups. There appears to be a dose-related shift in the first and fourth components of the vector (one increasing one decreasing) but not in the other three components. A description of the scatter of the points in five-dimensional space is supplied by the *variance—covariance matrix*, which is analogous to the variance encountered in Chapter 1. Analogous to the Kruskal—Wallis ANOVA of Chapter 6, there is a multivariate version known as *multivariate analysis of variance, or MANOVA*. Most standard statistical computer packages such as BMDP and SAS can handle sets of vectors in MANOVAs.

Once a significant dose-related shift in the means of these vectors has been determined, a biological description of its implications may prove difficult to make. The variance—covariance matrix yields a set of correlations among the components of the vector. These, combined with the degree of shift, provide a description of what happened. Interpretation of this type of data requires that the toxicologist think in terms of related organs and lesions. This section has been devoted to a discussion of how one might set up a multivariate vector of scaled response and how to test whether such vectors differ for treatment groups. Later we will discuss statistical techniques for interpreting differences in multivariate vectors.

Clustering Techniques

Figure 1 displays a typical result of a chronic toxicity study. The fomat is based on rectangular graph paper. Each row represents a specific lesion, such as hepatic tumor, nephritis, gastric inflammation, and lymphoma; each column represents a single animal. The rows have been arranged as a result of a clustering program which gathered together those lesions that tended to predominate in a given treatment. There is a mark on the graph paper made for each time an animal had a given lesion. The pattern of marks can be seen to cluster around some lesions for the controls and then move toward other lesions as the dose increases.

This is the idea behind clustering the results of a chronic toxicity study into collections of related lesions. It leads to a dose response that is difficult to describe in terms of some increasing mathematical function but which is apparent to the eye once the clusters have been identified.

Table 2 Lesions in Male Mice Found at Terminal Sacrifice
Summary of Statistics for Scaled Lesions

	Site of lesion				
	Bladder	Kidney	Liver	Lymph	Other
Mean vector					
Controls	0.60	0.95	2.75	2.30	2.45
20% Xylitol	2.20	0.85	2.45	0.65	2.05
Variance—covariance matrix					
Bladder	1.96	−.23	−.36	0.23	1.04
Kidney	−.23	1.08	0.60	1.02	−.62
Liver	−.36	0.60	2.58	0.90	0.13
Lymph	0.23	1.02	0.90	2.34	0.18
Other	1.04	−.63	0.13	0.18	2.79
Correlation matrix					
Bladder	1.00	−.16	−.16	0.11	0.45
Kidney	−.16	1.00	0.36	0.64	−.36
Liver	−.16	0.36	1.00	0.37	0.05
Lymph	0.11	0.64	0.37	1.00	0.07
Other	0.45	−.36	0.05	0.07	1.00

The concept of a cluster of lesions is similar to the medical idea of a syndrome. A physician determines that a patient's disease course follows that of some established syndrome if the patient appears to have most (but not necessarily all) of the symptoms associated with the syndrome and few (not necessarily none) of the symptoms that would be antithetical to the syndrome.

For ease of interpretation, it would be best to define the clusters of lesions in terms of some a priori biological hypothesis, and eventually it should be possible to modify clusters found empirically by statistical methods to reflect what is known about the interrelationships of organs and organ systems. However,

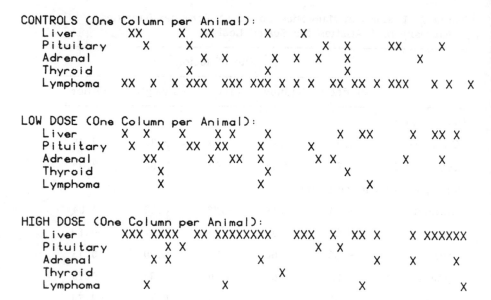

Figure 1   Patterns of tumors in male rats (NTP Study #226).

until a firm set of biologically defined clusters is agreed upon, there are computer programs that can be used to produce suggested clusters for any given study.

We can list the entire set of organs examined on each animal, changes in blood chemistry and weights, and numerical observations from gross pathology, indicating either a severity scale or a 0 (no lesion) or 1 (lesion) for each of the organs. Suppose there is a list of 40 such items for each animal. Then any two animals will have a "distance" between them defined by the lack of similarity of these 40 numbers. The distance function can be based upon the sum of the squared deviations or upon some probability-based function such as *Kullback's information function* (Kullback, 1968). The usual computer cluster programs then proceed in one of two ways: Either they start with individual animals, cluster those pairs that are closest together, and then go on to create larger clusters from those pairs of clusters that are closest together, and so forth, or they can start with the entire set of animals as one cluster and separate it into a small number of clusters whose elements are closer to each other than they are to elements of other clusters, further subdividing the first clusters into smaller clusters, and so on.

The resulting clusters are usually displayed in terms of graphs, connecting the hierarchies of clusters to each other and identifying the elements of each cluster. The user of the program then chooses a hierarchical level where he believes the animals have been separated out in a sensible fashion.

Once clusters have been identified, the number of animals in each cluster can be determined for each dose, leading to a pattern as shown in Table 3. This can be viewed as a contingency table (albeit with more than two rows), and we can test the hypothesis that the treatment (columns) are independent of the clusters (rows). If we reject this hypothesis, then we can say that the treatment had some influence on the syndromes of senile lesions. There are techniques for subtable analysis (most of which are available on large computer statistical routine packages like SPSS, SAS, and GLIM) that will enable the toxicologist to identify the clusters that are affected and to what extent.

One additional advantage of reducing the lesions to clusters is that it enables us to make one single hypothesis test of treatment effect that includes all the observed lesions, without worrying about adjusting for multiple tests or about whether the data have been cut up into too many small, uninformative pieces. Yet, by including all subtle differences of lesion in the initial listing per animal, we can retain the ability to see if the treatments had effects that were different for different types of tissues in the same organ.

If the treatment has induced changes in lesions, then the clusters will tend to identify patterns of lesion related to these changes, since dosed animals and controls will tend to have different lesion patterns. Thus the clustering of lesions is a sort of self-correcting procedure which takes advantage of the natural fall of the data to produce powerful tests of effect.

Table 3 describes the same set of male mice in the study comparing controls to three levels of xylitol as was discussed in the previous section. A comparison of Tables 1–3 shows how the same set of data can be examined using two different statistical techniques. The use of multivariate vectors lead to a graded numerical description of the degree of change (the mean levels of the elements of the vector) and the interrelationships (the correlations). Clustering allows us to use a great deal more information on each animal, but it produces a more complex description of the study.

Both these techniques, however, begin to describe more closely what actually happened in a chronic toxicity study.

Table 3 Clusters of Senile Lesions Associated with Significant Differences Among Treatments for Male Mice

Description of clusters	Counts of animals					Significance levels	
	Controls	Sucrose 20%	Xylitol 2%	10%	20%	Xylitol dose response	Sucrose vs controls
Death before 75 weeks with bladder tumors or hyperplasia but without lymphoreticular or kidney tumors	0	2	2	11	11	<.0001	.0825
Death after 74 weeks or terminal sacrifice with bladder tumor or hyperplasia but without kidney tumor	3	5	4	18	27	<.0001	.2566
Death with liver tumor or hyperplasia but without tumors in lymph, lung, kidney, or bladder	18	10	10	2	4	>.9999	.9646
Lung tumors without tumors in lymph, kidney, or bladder	13	19	18	5	7	.9949	.1519
Total number of animals examined	79	83	75	86	86		

There are well-developed statistical techniques for handling this complexity of data. Just as later we will examine some of these techniques for multivariate scaled vectors, we will also describe techniques for dealing with categorized data and the resulting clusters.

Bayesian Analysis

Bayesian statistics is an area of great promise but with little application in toxicology so far. Once we have proposed a probability distribution for the data observed in a study, then the whole of statistical analysis can be summed up as an investigation of the values of the unknown parameters of that distribution. If we are counting the number of animals with a specific lesion and modeling these counts as binomial variables, then we use statistical techniques to examine the underlying p values. If we are measuring changes in blood chemistry and model the data as arising from a Gaussian distribution, we use statistical techniques to examine the means and variances.

The basic idea behind Bayesian statistics is that the parameters of these models are themselves random variables; that is, they change from study to study in a random fashion. If this is true, then we may not want to know the exact value that a parameter took on for a given study but wish to characterize the probability distribution of that parameter across all studies.

A central idea of Bayesian statistics is that we arrive before the study has begun with a previously developed probability distribution of the parameters. To fix ideas, let us suppose we are about to run a chronic toxicity study and are interested in the number of animals in each group with malignant tumors. The underlying p for the control group has a probability distribution and we have a *prior* estimate of that distribution. We then run the study and observe the results. We use the data from this study, not necessarily to answer specific questions about the current value of the parameter but to adjust the prior estimate of the distribution. This adjusted estimate, which uses the observed data, is called the *posterior* distribution.

This sequence of estimates, from prior to data to posterior, may not seem appropriate to toxicological research at first glance. We are usually involved in determining if, under the conditions of study, the treatment had a toxic effect and, if so, to characterize that effect and its dose response. We concentrate on the current study and usually limit ourselves to

that study only; however, eventually the results of many studies will have to be combined so that society can decide how to
handle some particular substance. At this point the differences
between studies often add confusion rather than light.

For instance, one of the problems with the acute LD_{50} is that
determinations made in one lab often differ from those made in
another lab, sometimes without even overlapping confidence intervals. Suppose we assume that the probit model is correct
and we are seeking the two parameters, the LD_{50} and the slope
of the linear segment. If the estimates of these differ from lab
to lab, it might be possible to model this problem as if these
parameters were, in fact, random and dependent not only on
the compound being tested. If so, the probability distribution
of the individual study parameters might yield bounds for these
parameters that can be derived from the posterior distribution.
This combination of many such studies has already been suggested in the statistical literature but does not seem to have
been actually done for any particular compound.

Is there a similar problem for chronic and subchronic studies? For toxicity studies in drug research we usually seek a
starting dose in humans and use the minimum toxic and no-effect
doses from subchronic studies run on different species to predict a safe starting dose. Bayesian techniques might be useful
there; however, the current methods appear to be quite successful without bringing in more statistical jargon.

The results of chronic studies are often used to determine a
reasonable environmental level for a compound shown to be toxic at high doses. This might be a place to apply Bayesian statistics, and methods have been suggested by researchers interested in risk analysis. However, the usual problem in risk
analysis is adopting a small number of rodent studies done at
high doses to estimate risk associated with human exposure as
very low doses. This seems to be a problem beyond the usual
formulation of Bayesian statistics.

The fact that most chronic studies are done as part of a
long sequence in the same laboratory suggests that Bayesian
statistics should be of some use. We have large amounts of the
type of prior data that the Bayesians need. The problem is apparently one of finding an appropriate toxicological question.
If we use this prior information and the current study to compute a posterior distribution for the control probability of lesion
occurrence, we might be able to test the treatment incidence of
lesions against the most probable value of the control probability. This is what has been proposed by Tarone (1982) and

Dempster et al. (1983), among others. However, the question then being posed is whether the treatment incidence is within the random pattern seen in the past among the controls. Since the laboratory conditions change from study to study, it is conceivable that the current study is "unusual" but provides strong evidence of a treatment effect under those conditions.

Which question, then, should be answered? (1) Is there a treatment effect under current conditions? or (2) is the treatment incidence within what would normally be seen among controls in this lab? If it is the first, there is no apparent place for Bayesian methods; only for the second question do these methods seem appropriate.

Measures of Risk

Traditionally, toxicology has been concerned with the interpretation of individual animal studies, and only peripherally with the extrapolation of those studies to human risk. Prior to the 1960s the most extensive attempts to synthesize material from different studies were run within the drug industry, where subchronic studies run on several species were used to estimate an entering "safe dose" for human trials. For food additives or industrial exposure, safe doses estimated from animal studies were converted to bounds on human risk by dividing them by appropriate safety factors (usually 100–1000 for food additives).

The 1960s and 1970s saw the development of large chronic studies in rodents, prompted primarily by the fear of latent irreversible lesions (such as cancer) that might occur from low-level exposures to certain chemicals. From the beginning, these studies were intended to be extrapolated to human risk. First thoughts were that a small number of dangerous substances would be identified by such studies and then eliminated from the human environment or brought under strict control. If there were only a small number of such compounds and if very low levels could be detected but not necessarily controlled, Nathan Mantel of the National Cancer Institute proposed a mathematical model for estimating the risk that might come from such exposure (Mantel et al., 1975).

He proposed a simple mathematical model that projected risk in terms of the probability of lesion occurrence. (Since cancer was the major lesion of interest, the rest of this discussion will use that word.) He assumed that the probability of cancer was an increasing function of dose and of dose only:

P(cancer) = f(dose)

He also proposed that society set a low probability as "safe"
(his first paper suggested 10^{-9}). We would find a convenient
function f that was reasonably sure to produce probabilities
that were higher than the true probability. Inverting the function, the dose associated with this low probability would then
be found. This dose would be an upper bound on the true
dose that might induce such a low probability.

Since then, there has developed a large literature on estimating risk in this fashion. Mantel first proposed the probit as
a convenient bounding function for f( ). This was rejected by
the majority of articles in favor of more conservative bounding
functions, functions based upon fitting the observed data, or
functions based upon a simplified mathematical model of carcino-
genesis. All of these techniques for estimating risk violate a
fundamental axiom of applied statistics:

> Mathematical models should be used to project results
> only within the range of the observed data.

This axiom is based upon the realization that all mathematical
models are poor approximations of reality. As long as observed
data are used to correct and adjust the parameters of the model,
it has been found that different models predict much the same
results. As soon as models are extended beyond the range of
observed data, however, the arbitrary components of the model
begin to dominate, and models that all fit the observed data
reasonably well will predict drastically different results outside
that range.

Thus it would make sense within traditional applied statistics
to model the probability of cancer for doses near those used in
the chronic studies, but no well-trained statistician would attempt to estimate probabilities of cancer associated with doses
well below or well above these. The main argument of those
who do this is that the models they propose are not meant to
"fit" the data; instead, they are bounds on all the models that
might be expected to be true. Figure 2 displays this argument
graphically. It is claimed that all dose response models are
curved in the low-dose region so that any straight line that
goes through zero will lie above some portion of the curve.
Many proposals are to take a point in the observed dose range
that is sure to lie above the dose response curve and draw a

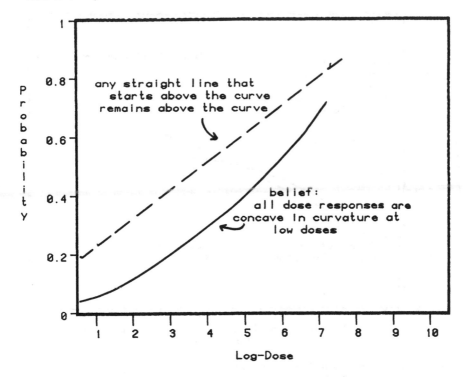

Figure 2 Linear "bounds" in low-dose extrapolation.

straight line back from there, using the dose associated with a low probability on that line as a *virtually safe dose* (or VSD).

There are problems with this simplistic view. First of all, not all dose response patterns are bounded by a straight line. Some models will curve the other way, as in Figure 3. In fact, analysis of the incidence of hepatic tumors induced by vinyl chloride in rats indicates that such a "supralinear" probability curve fits those data. If we consider the class of all possible dose responses, then there is a reasonable one that can never be bounded above by a straight line. Suppose there is a small subset of the population that is susceptible to a single molecule of the carcinogen. Then there will be a jump in the probability of cancer just beyond zero dose, as indicated in Figure 2, and any straight line will fall below the curve at that point.

This whole argument resembles medieval scholasticism: it can be discussed and argued without ever looking at experimental

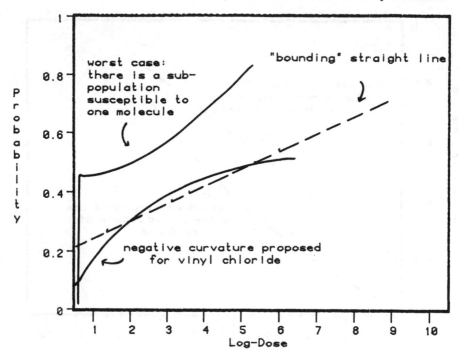

Figure 3   Supralinear low-dose patterns.

results (and, unfortunately, it often is). Its main stock in
trade is the sophistry of dreaming up ever more sophisticated
structures for the probability function f(dose). Since Mantel's
original proposal is without precedent in the statistical literature
and since it violates the principles of applied statistics, the
statisticians who argue over these arcane models are as much
strangers to the arguments as the toxicologists who try to follow them.

Anyone who has looked at the data of a chronic study will
recognize that there are serious faults in Mantel's original
mathematical model. It is not possible to model the probability
of cancer as a function of dose only. Many compounds that
have been called carcinogens in these studies produce multiple
patterns of lesions. For instance, vinyl chloride produces lethal
zymbal gland tumors early on and liver and brain tumors in animals that survive to old age. Each of these tumor incidences
produces a different estimate of the virtually safe dose.

Other compounds involve natural background tumors (such as mouse hepatic tumors), and it becomes necessary to include a factor for background incidence in the function f. Finally, the duration of and time pattern of exposure play a role. A large study in mice run on the carcinogen AAF, which included groups of animals withdrawn from treatment after 12, 18, and 24 months of continuous exposure, suggested that the initial lesion that leads to cancer is reversible, so it might be that, when time is brought into the picture, the function f may even turn downward.

Thus examination of the available data suggests that, at a minimum, risk must be expressed so that

P(cancer) = f(dose, time, lesion, species, background)

Robert Seilkin (ED-01 Task Force, 1981) has proposed that the first two arguments be used to construct the probability of cancer as a function of time and dose. Then the measure of risk would no longer be the probability of cancer but some measure of time associated with the occurrence of cancer. He proposed two such measures (although others can be derived by the same line of reasoning). One of these considers the average shortening of time to tumor development that occurs at a given dose (assuming a background incidence). He suggested that a virtually safe dose is one that shortens that average time by less than some small factor. Another measure of risk Seilkin proposed compares the average time to tumor development against the expected life span of the species. In general, Seilkin's measures of risk assume that any animal, if it lives long enough, will get cancer. He would then have us estimate the degree to which a given dose of the suspect compound will modify that event.

One advantage of Seilkin's measures of risk is that they project VSD doses that can be tried out in feasible experiments and they project events at those doses that are observable with only a few hundred animals. In this way predictions based upon Seilkin's measures can be tested in an experimental setting. There is no way of testing a prediction that the incidence of tumor is less than 10^{-9} without using billions of test animals.

If measures of risk are to be based upon what is actually observed in a chronic toxicity study, then it will be necessary to bring still more arguments to the function f, including the multiplicity of lesions and the grading of precursor lesions. As of

this writing (1986), a great deal of statistical work remains to
be done before the interpretation of chronic toxicity studies
reaches the level of usefulness developed for subchronic studies
prior to 1960.

Estimating Effects in a Multivariate Setting

Earlier we examined how the multitude of lesions from a chronic
study can be reduced to vectors of scaled numbers, each element of the vector describing a degree of derangement in a single organ or organ system. Once we set up such a multivariate
description of the study, we can test whether treatments are
significantly different from controls by using multivariate ver-
sions of the univariate statistical tests described in Chapter 8.

What happens if there is a "significant" effect? It then becomes necessary to interpret that significance. This section
deals with interpreting multivariate statistics.

Multivariate statistics can be interpreted either in terms of
the algebraic symbolism used to calculate effects or in terms of
multidimensional geometry. Neither is easy to understand. In
the algebraic symbolism we usually look upon a vector as a set
of numbers arranged in a vertical column,

and we work with linear combinations of the elements of a vector; that is, we turn the column of numbers into a single number by finding another column of numbers, say,

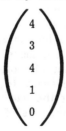

and adding up the piecewise products of the two columns, as follows:

$$(3 \times 4) + (2 \times 3) + (0 \times 4) + (1 \times 1) + (1 \times 0) = 19$$

This is a purely abstract tool. It gets used over and over again in multivariate statistics for a large number of different purposes. In a toxicological context, the first vector might describe the scaled lesions in five organs for a particular animal and the second vector might be a set of relative weights which describe the importance of each organ in determining the toxic effect. Then this *linear combination* provides a single number to describe the toxicity of treatment in this animal for this particular effect.

Or the two vectors might represent two different animals. Then the linear combination tells us something about how similar the lesions are in the two animals. If both animals had the same degree of lesion in the same organs, the linear combination will be large in value; if they tend to differ in their patterns of lesions, several components of the vector for one or the other animal would be a zero, and the linear combination will tend to be small in value.

Although there are other statistical uses for linear combinations, these two are the most important for toxicology. We can use linear combinations to reduce a vector of lesions to a single number representing an effect or to compare the degree of similarity between two vectors of lesions. It can be seen that both techniques reduce the vectors to something understandable but lose some information in the process. One particular set of weights will produce a single number to describe a toxic effect, but another set of weights will give another number.

The second method of interpreting multivariate statistics involves the use of geometry. No one can visualize more than three dimensions, so, although we will usually be dealing with much higher dimensionality, the terminology (and the picture we are asked to keep in mind) is based entirely on three-dimensional geometry. The sheets of paper used in this book are two dimensional and the reader is required to "see" three dimensions in the drawings that appear here. This in itself is often a barrier to understanding. As an aid in seeing the three dimensions, it is a good idea for the reader to focus on one corner of the room he is sitting in. The vertical edge of the room is one axis; one edge running at right angles to this along the floor is another axis; the other edge running at right angles to

both of these is the third axis. The room in is one octant of
the three-dimensional space defined this way. There are three
other "rooms" on the same floor that describe the other octants
where the distance along the vertical axis is positive and four
"rooms" on the floor below for the remaining octants, where the
distance along the vertical axis is negative.

A point in this 3-space is equivalent to a 3-vector, such as

$$\begin{pmatrix} 2 \\ 1 \\ 5 \end{pmatrix}$$

by calling the axis along the floor to the left the first axis and
measuring 2 units along that, calling the other axis along the
floor the second axis and measuring 1 unit along there, and,
finally, measuring 5 units up the vertical axis. These three
distances define a single point in the room as in Figure 4. If
one were to scale the lesions in only three organs, the 50 animals in a single group could be represented as a scatter of 50
points.

Imagine these 50 points as tiny pinpoints of light, shinning
in all directions. You would see a spherical-shaped scatter of
lights, with many of them clustered near the center of the football and shining together the brightest, fewer and fewer close
together as you get further from the center, and a few scattered, far-off ones a great distance from the better-defined region of the sphere of light. Furthermore, the lights would be
projected onto one of the axes, say, the vertical one, and you
would see a bright region of light projected by the center of
the sphere and less and less light on the axis away from that
center. The sphere might be twisted in such a way that the
brightest projection on the axis results not from the center but
from the way in which a number of peripheral lights line up.
We could also draw an arbitrary line from the corner of the
room out at an angle and examine the projection of light along
that line.

The projection of light from the sphere onto a line drawn
from the corner of the room describes the first type of linear
combination. The distance along that line to the point where
the projection is brightest is the number we would get by computing the linear combination of the vector of average values
and the three direction numbers that define that line. Thus a

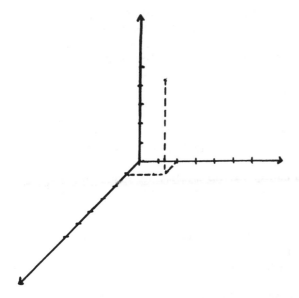

Figure 4 A point in three-dimensional space.

weighting scheme that lets us reduce a multivariate vector to a single number can be thought of as a single line projecting out into space from the origin (the corner of the room).

When we use linear combinations to compare two vectors, we think of the geometry a little differently. Suppose we had two points defined in the room. We construct a line from the origin to each point, so the two lines meet at the origin. The linear combination of the two vectors is directly related to the cosine of the angle between these two lines. If the two lines are perpendicular to each other, the linear combination is zero. From this, we get the geometric interpretation that two vectors that are completely different from each other (having a zero linear combination) are *orthogonal* to each other. In fact, the concept of correlation is tied directly to the angle between two vectors. The Pearson product—moment correlation is the cosine of the angle between the two sets of numbers being compared, and random vectors that are orthogonal to each other are statistically independent of each other, since they have zero correlation.

In interpretations of analysis of variance, we often compute "orthogonal" contrasts because statistical significance found with

respect to one contrast is independent of statistical significance
found with a contrast orthogonal to it. The terminology and the
geometry are all linked. If the reader can cultivate the ability
to think in three dimensions and then make the leap of faith that
says that such pictures hold true for higher dimensions, a great
deal of statisitical theory will appear as a unified and sensible
body of knowledge.

Using either or both interpretations (the algebraic one or
the geometric one), we can now examine the situations where
the multivariate scatter of points resulting from two or more
treatments differ significantly. When we say that two collections
of vectors differ significantly in the means, we are really saying
that there is at least one linear combination of the means with
the following property:

> If we had computed that linear combination for each
> animal and run a univariate analysis of variance on those
> numbers, we would have gotten statistical significance.

In the example of xylitol treatment given in earlier, application of multivariate analysis of variance shows that the mean
vectors differ significantly between the controls and the xylitol-
treated mice. The major difference occurs between the controls
and the two higher doses. Recall that the vector of means for
the controls was

$$\begin{pmatrix} 0.60 \\ 0.95 \\ 2.75 \\ 2.30 \\ 2.45 \end{pmatrix}$$

and for 20% xylitol it was

$$\begin{pmatrix} 2.20 \\ 0.85 \\ 2.45 \\ 0.65 \\ 2.05 \end{pmatrix}$$

The five "organs" involved were

Bladder
Kidney
Liver
Lymphatic system
All others

suggesting that the "effect" involved an increase in the severity
of lesion for the bladder and kidney and a decrease for the
others. We can use statistical techniques to find a linear combination that will be significant by itself, and we can then examine how the resulting weights might make biological sense.

More on Linear Combinations

Since it is so difficult to interpret a vector of numbers, most
multivariate analyses end up studying linear combinations of the
means. In the previous section we suggested that one can derive a most significant linear combination from the data after
finding that there is a significant difference among treatments.
In areas of research such as sociology and psychology, where
multivariate statistics are widely used, this is not the usual
way, however.

One method, used a great deal in medicine, is to derive a
"sensible" linear combination that represents what is known
about the biology being examined. For instance, rheumatolo-
gists assess the amount of swelling and pain patients have in
different joints, the degree of morning stiffness as reported by
the patient, and several identifiers of rheumatoid arthritis found
in the blood. From these, they have derived ad hoc indices
that measure the degree of overall illness. Ad hoc linear combinations of measurements have the apparent advantage that
they are based upon a careful consideration of the relative biological importance of the measures. In fact, they are often
based upon the general impression of experts of the relative importance of these measures as they have seen in the past when
describing patients.

Psychologists (and psychiatrists) often evaluate patients in
a multivariate fashion. For instance, there is a standard measurement scale for psychotic patients call the Brief Psychiatric
Rating Scale (BPRS). The evaluator examines 18 aspects of the
patient's behavior and rates each one on a scale of 1 (none ob-

observed) to 7 (severe involvement). The aspects of behavior include specific symptoms of mental disease, like hallucinations, grandiosity of manner, somatic concern, and blunted affect. When the scale was first used, psychiatrists constructed different sets of weights depending upon different diagnostic categories. For instance, if the patient were suffering from psychotic depression, those parts of the scale that dealt with symptoms of depression were weighted with a value of 2 and other symptoms were given a value of 1 or 0, depending upon how well correlated they seemed to be with depression.

Later, when a great deal of data had been accumulated, the BPRS underwent a *factor analysis* (Morrison, 1967). The idea of factor analysis is to examine the pattern of scatter observed and find some linear combination or combinations that reflect that scatter. In its most theoretical form, we think of the vector to be examined as having been generated by a small number of inherent "factors" plus a little random noise. The inherent factors are the truly independent underlying elements of the disease. If, for instance, we had measured 18 aspects of behavior, there might be only 5 fundamental components of personality disorder that lead to this psychotic behavior. We would seek five linear combinations of the observed data that are orthogonal and thus independent of each other and which "explain" the observations.

As originally proposed in the 1930s, factor analysis was not well defined and different schools of psychology developed different methods of estimating these underlying factors from the observed data. Some of this confusion still exists, but, almost by default, there is a standard method now widely used which is based upon the first successful attempt to estimate factors on the computer in the early 1960s. This method starts with a related but well-defined statistical concept, that of *principal components* (Morrison, 1967).

Going back to the football of tiny lights that represents the scatter of vectors of data in a geometric model, we can think of the football as having different axes. In three dimensions there are three such axes. The first axis goes through the center along the longest dimension of the football. If the football is flattened on one side, the second axis goes through the center, perpendicular to the first axis along the next longest dimension, and the third axis runs perpendicular to these two along the shortest direction (the flattened side). If this last axis is very short, we can think of the football as being essentially a two-dimensional ellipse lying at an angle in space with a small scatter of points of the plane of the ellipse. The planets around our sun, for instance, line up along a flat elliptical plane, with some of the moons of the larger planets and the furthest planet, Pluto, orbiting just slightly outside that plane.

In higher dimensions we can think of the various axes of the multidimensional "football" in the same way. In fact, with a conceptually simple but time-consuming series of calculations, we can find the linear combination of data that describes the longest axis. We then subtract that linear combination from all the data, thereby reducing the dimension of the football by one, and run the same algorithm again to get the second longest axis, subtract that, and so on. This is clearly something that can be done only on a computer, and so the calculation of these principal components had to await the arrival of modern computers.

Since the principal components are based upon the observed scatter of points, each of them "explains" some of the overall variability. We can, in fact, compute a generalized variance of the entire set of points and determine what percentage of that generalized variance is "explained" by a given component. In practice, we usually continue computing principal components until we have explained about 90% of the variance. In the analysis of the BPRS, for instance, well over 90% of the variance is explained by the first five principal components.

The principal components are not quite the factors of factor analysis, but they are a good starting point. We may not even need to go any further. For instance, it may turn out that almost all the variance is explained by only one or two components. If the first component explains most of the variance, the scatter of points is along a line in space, and the lesions found in specific organs are so highly correlated that any one or the average of all of them is adequate to describe what has happened. When only two components explain most of the variability, it often happens that the first component is the average of all the elements and the second component is a weighted difference, contrasting one set of lesions to another.

If more than two principal components are needed to explain most of the variance, we can then move into a more structured factor analysis. We first have to assume that we know the number of factors involved. We have to pick a specific number so we can construct the projections of the principal components on that smaller number of vectors. Suppose we decide that there are five factors. The computer programs now available start with the first five principal components. By construction,

these are orthogonal to each other. We then rotate this system
of orthogonal vectors. It is as if we were to rotate the corner
of a room in one, two, or three directions. The five linear
combinations will change together as the system is rotated, but
they will remain perpendicular to each other (so that any two
linear combinations will add up to zero). As they rotate, we
compute a measure of the uniqueness of the vectors, called the
communality. We are looking for a set of factors that are not
only independent of each other but that are weighted almost entirely, each one, on a small number of elements of the vector.

In the example of the BPRS, the first five principal components were rotated until each one had most of the weight on
only five or six items. The items that were heavily weighted in
each factor were then used to name the factors.

In a chronic toxicity study we might hope that a factor
analysis of lesions will enable us to find groups of lesions that
are related and which might be expected to increase together
under the influence of a particular type of toxic activity.

Yet another application of multivariate statistics to the BPRS
offers a possibility for toxicology. Patients were treated with
different classes of drugs and the changes in BPRS were noted.
Once a large number of such treatments had taken place, it was
possible to obtain a prototypical pattern of change for benzodi-
azepams, phenothiazines, and so on. Suppose a new drug has
been developed. It is used on a group of patients and the
average change in BPRS items is computed. The correlation
between this average change and the average changes associated
with established classes of drugs is then computed, and the new
drug is classified in terms of the closest prototypical vector.

Such a procedure could be used in toxicology. We could examine the chronic dosing studies for compounds with well-
established patterns of toxicity. The new study could then be
compared in its mean vectors of effect to establish the general
nature of toxicity of the test compound in terms of known compounds and also to compute a relative potency comparison.

Clustering and Directed Graphs

One of the problems with the use of multivariate vectors is that
we have to start with a scaling of lesions within organs. To
scale lesions we not only have to order them in severity but also adopt a convention about degree of difference. When we
scale hyperplasia as 1, benign tumors as 2, and malignant

tumors as 3, we are assuming that the "distance" between successive lesions is the same. When we look at average changes, a move from hyperplasia to benign tumor has the same value as a move from benign to malignant tumor. The average of changes for individual animals then mixes up the changes between different lesions and we can no longer make a distinction between them.

We gain a great deal of statistical machinery from that scaling, and attempts to test for treatment-related differences or to estimate the degree of effect become much more difficult without scaling. If we do not scale, we can still attempt this, but the mathematics are more complicated and the results of those calculations will be even further removed from straightforward interpretation.

In an earlier section we looked at the clustering of animals in terms of patterns of lesions. We can look at these patterns in terms of a multidimensional contingency table. Each dimension represents a different organ. Each organ is classified by the type of lesion seen in it, so two dimensions of the table might compare the incidences of animals that have liver and lymphatic system lesions together, giving counts of animals with both liver and lymphatic system malignant tumors, those with liver benign tumors only, and so on. This is expanded into higher dimensions, where liver, lymphatic system, and kidney are compared, and still higher, where four or five organs are combined, up to a full table where all 50 organs usually examined are compared across and between organs, with another dimension for treatment, still another one for time of study, and still others for the initial conditions of the animals, sex, and so on.

In the past 10 years an extensive amount of theoretical work have appeared in the statistical journals dealing with this sort of a problem, a very large multidimensional table of interrelated events. The basic method of analysis is to assume that the probability of finding an animal in a particular cell of the table (malignant lymphoma, no liver lesion, bladder hyperplasia, female, control, final sacrifice, etc.) has a general mathematical form, so

P(cell) = f(treatment, organs, sex, etc.)

When the function is based upon a linear model for the logarithm of the probability, it is called a *log-linear* model; when it is based upon a logistic function, it is called *logistic regression*.

Other models exist, but these two are the ones for which
standard statistical packages like SAS and BMDP now have the
ability to run complete analyses. The proponents of these new
techniques claim that these are very general tools that can be
adapted to almost any purpose.

It might be interesting, however, to examine a particular aspect of log-linear models as they are used in econometrics. It
does little for comprehension to draw up a complicated contingency table with all the cross-classifications involved. Instead,
the econometrician thinks of the problem in terms of *graph
theory* (Wermuth and Lauritzen, 1983). In this model we think
of the connections among the various components of the table in
terms of a tree of relationships. We need a picture here to aid
our thinking; its simplicity will enable us to concentrate on a
small number of aspects of the problem but will also hide some
assumptions that may have to be explored.

Figure 5 is a graph of the results of a chronic toxicity trial.
The graph consists of a set of points connnected by lines.
Each "point" on the graph is a particular dimension of the table.
One point represents the treatment, another point the lesions
seen in the liver, another the lesions in the lymphatic system,
another the sex of the animal, and so on. When a line connects
two points, it means that some sort of a relationship is expected
between these two dimensions. Two points may be connected
directly, indicating a relationship between them that cannot be
entirely by other dimensions. Two points may be completely

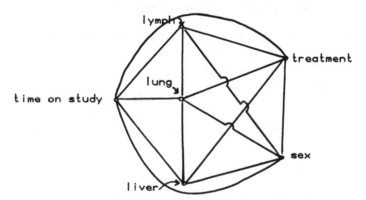

Figure 5   Graph of multidimensional contingency table. A fully
saturated model with all nodes connected.

isolated with no lines connecting them. This indicates that they are completely unrelated. Finally, two points may be connected but only through a path that must pass through a third point lying between them. This indicates that their relationship can be explained completely by the relationship they have to other parts of the table.

Figure 5 shows a graph where all points are connected directly. Figure 6 shows one where the points representing lesions in specific organs are connected only through treatment. The schema described in Figure 5 indicates that all lesions are interrelated and that whatever effect is related to treatment is confused by positive and negative correlations between lesions in different organs. Figure 6 indicates that the various organs are conditionally independent. If the occurrence of lesions in two organs is correlated, it is entirely due to the effects of treatment. Figure 6 describes the naive view that we can examine the effects of treatment on individual lesions in isolation and expect to determine toxic effects. Figure 5 describes the pessimistic view that it is impossible to examine lesions in isolation and that effects of treatment can only be seen in terms of the entire animal.

The theoretical ideas involved here are *statistical independence* and *conditional independence*. The occurrences of lesions in two organs are statistically independent if the probability of lesion in one organ is unaffected by the occurrence or nonoc-

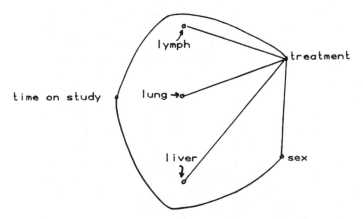

Figure 6 Graph of multidimensional contingency table. Lesions are independent and conditional on treatment.

currence of lesion in the other organ. They are independent
conditional on treatment if the probability of lesion in one organ
is unaffected by the occurrence of a lesion in the other organ,
once we account for the effect of treatment on both organs si-
multaneously. If lesions occur independently in two organs,
then it is possible to estimate the probability of lesion in each
organ as a function of the dose of toxin, and it is possible to
test for a significant treatment effect separately on each organ.
Even if the occurrence of lesion in one organ makes it impossible to observe a possible lesion in the other organ, we can
estimate effects and test for them as long as there is statistical
independence.

For instance, suppose that the animals are subject to malignant lymphomas early in the study, so that a number of the
animals die with the lymphoma and never have a chance to de-
velope later-occurring lesions or the invasive lymphomas make
it impossible to examine the bladder and other small organs for
lesions. If the occurrence of malignant lymphomas is independent of the occurrence of these other lesions (even if the independence is conditional on treatment), then the statistical
methods used to adjust for early deaths and time on study are
all valid for purposes of determining treatment effects. Furthermore, it is theoretically possible to estimate the dose response
associated with different treatments.

However, if the occurrence of lymphoma is not independent
of the occurrence of some other lesion (such as a hepatoma),
then it is impossible to construct a valid test of the hypothesis
that the treatment induced lymphomas (or hepatomas), and it is
impossible to construct a consistent estimator of dose response.
The entire collection of dependent lesions must be considered
as a group. It is possible, of course, to make some sort of
analysis. It may turn out that there are clusters of lesions,
with lesions within a cluster being statistically dependent but
with each cluster statistically independent of the other—a situation explored in the earlier section on clustering techniques.
Then we could examine if the treatment has an effect on a giv-
en cluster. In addition, if we could order the elements of a
given cluster so that one set of lesions is considered "worse"
than another, we might even be able to estimate dose responses.

The situation where there are dependent clusters of lesions
can be represented by Figure 7. Here the graph shows arcs
connecting treatment to each organ, but additional arcs connect certain organs together. Those clusters of organs which

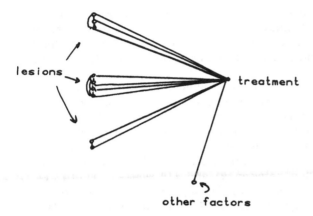

Figure 7 Graph of multidimensional contingency table. This figure shows independent clusters of dependent lesions.

are connected only by way of the treatment node are independent (conditional on treatment) and each cluster can be examined for treatment effects.

How can we know whether individual organs have lesions independently of other organs or whether there are clusters that are independent of each other? There is a statistical test for conditional independence based on the log-linear model. We start with a log-linear model as depicted in Figure 5, where all nodes are connected. If there is no direct arc between two organs, then all possible interrelationships between those two organs have to include an element from the set of nodes that do connect them. So a log-linear model is constructed that includes only these types of relationships. The fit of the data to the two models is compared. A test statistic is constructed that is used to reject the hypothesis of conditional independence (the second model) in favor of the first. This is done for all pairings of organs, producing a graph similar to that in Figure 7.

A model like that in Figure 7 leads to difficulties in interpretation. Suppose, for instance, that the dependent organs include the lymphatic system and the liver and that the "effect" of treatment is to decrease the incidence of lymphoma while increasing the incidence of hepatomas (which is what apparently happened to the sucrose "control" in the xylitol study used for illustration in earlier sections). How can this be interpreted?

An answer to this question is outside the realm of statistical analysis. Statistical techniques can be used to identify more or less reasonable mathematical models to describe the data at hand; it is up to the toxicologist to make biological sense out of those models.

Two sets of data can be used to illustrate the ideas of this section. Table 4 displays the incidence of lymphomas and hepatic tumors in male mice subjected to DDT taken from an article by Wharendorf (1983). The mice have been divided into two groups, one consisting of animals that died or were killed between 35 and 90 weeks and the second consisting of animals that died or were killed between 91 and 110 weeks. There are three treatments: controls, mice fed high doses of DDT for 15 weeks, and mice fed high doses of DDT for 30 weeks. Table 5 displays a similar set of data for mice treated with four doses of xylitol (and controls). More organs than just the liver and the lymphatic system were examined in both studies, and the information available provides for a finer gradation of lesions; however, it is conceptually easier to keep only two lesion types in mind for the analysis that follows.

Figure 8 displays the fully saturated graph, where both lesion types, both sacrifice groups, and all treatments are connected in every possible way. Figure 9 displays the graph where the occurrences of lymphoma and hepatic tumor are connected only through the treatment and whether the animal reached final sacrifice. If the fully saturated graph (Figure 8) is true, the model will predict the number of animals actually observed; that is, a fully saturated model is one that describes exactly what was seen. If the conditionally independent graph (Figure 9) is true, Table 6 displays the numbers of animals that would have been expected in the DDT study, and Table 7 displays expected numbers in the xylitol study.

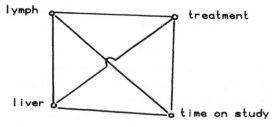

Figure 8   Fully saturated model xylitol and DDT studies.

Table 4 Lymphomas versus Liver Tumors—DDT Study

Lymph	Liver	Group I-3			Group I-4		
		Controls	15-week	30-week	Controls	15-week	30-week
−	−	74	51	24	45	17	7
−	+	14	20	52	36	27	46
+	−	26	18	8	22	11	3
+	+	1	1	4	10	9	9

Test of independence, conditional on treatment: $p < .01$.
Source: Data taken from J. Wahrendorf, JNCI, 70, pp. 915–921, 1983.

Table 5 Lymphomas versus Liver Tumors—Xylitol Study

Lymph	Liver	Group I surviving less than 91 weeks: xylitol as % of feed				Group II terminal sacrifice: xylitol as % of feed			
		0%	2%	10%	20%	0%	2%	10%	20%
−	−	27	33	31	31	15	21	10	17
−	+	17	9	10	7	23	19	8	13
+	−	4	6	8	5	7	5	2	6
+	+	3	0	1	1	3	4	2	4

Test of independence, conditional on treatment: $p > .50$.

   The expected numbers in the DDT study are significantly different from the observed numbers, suggesting that a conditionally independent model is not appropriate for use in analyzing these data. If we wish to run a formal hypothesis test of dose response, we have to consider both the occurrence of liver tumors and that of lymphoma as being related. This is a negative relationship and, as shown in Table 8, treatment induces a significant increase in liver tumors associated with a decrease in lymphoma.
   On the other hand, the expected numbers of animals in each group that are predicted by the conditionally independent model for the xylitol data are very close to those observed. We can

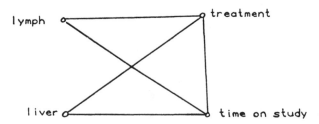

Figure 9 Conditionally independent model, xylitol and DDT studies.

Table 6 Expected Frequency of Lymphomas and Liver Tumors—DDT Study Expected frequency under model with incidence of tumor in lymph and liver independent, conditional on treatment

Lymph	Liver	Group I-3			Group I-4		
		Controls	15-week	30-week	Controls	15-week	30-week
−	−	84	54	28	48	19	8
−	+	12	17	48	33	25	45
+	−	24	15	4	19	9	2
+	+	3	4	8	13	11	10

Table 7 Expected Frequency of Lymphomas and Liver Tumors—Xylitol Study Expected frequency under model with incidence of tumor in lymph and liver independent, conditional on treatment

Lymph	Liver	Group I surviving less than 91 weeks: xylitol as % of feed				Group II terminal sacrifice: xylitol as % of feed			
		0%	2%	10%	20%	0%	2%	10%	20%
−	−	27	34	32	31	17	21	10	17
−	+	17	8	9	7	21	19	8	13
+	−	4	5	7	5	5	5	2	6
+	+	3	1	2	1	5	4	2	4

Table 8 Overall Dose Response Patterns in Male Mice—DDT Study

	Percent of animals with tumors		
	Controls	15-week	30-week
Lymphomas	25.9	25.3	15.7
Liver tumors	26.8	37.0	73.5

Table 9 Overall Dose Response Patterns in Male Mice—Xylitol
Study

Percent of xylitol in feed	Percent of animals with tumors			
	0%	2%	10%	20%
Lymphomas	17.2	10.3	18.1	19.0
Liver tumors	46.5	33.0	29.2	29.8

now examine the two lesions independently to determine whether
there was a treatment-induced effect. A formal hypothesis test
of dose response indicates a significant decrease in liver tumors
with increasing dose, and a separate test for dose response indicates no evidence of an effect on lymphomas (see Table 9).

At this stage the statitical analysis has offered as much in-
sight (or confusion?) as it can. We can conclude with some
degree of assurance that DDT has an effect on the syndrome of
lesions that includes a negative relationship between lymphomas
and hepatic tumors and that its effects on the liver cannot be
examined in isolation. We can also say with some degree of
assurance that xylitol has an effect on the incidence of liver
tumor, independent of what happened to the lymphomas. It is
left up to the poor toxicologist to supply a biological interpretation for these conclusions.

References

Armitage, P. 1955. Tests for linear trends in proportions and frequencies. *Biometrics* 11:375–386. The methods discussed in this paper had, in fact, been "invented" before and used by others, since one way of deriving Armitage's test is to treat the occurrence and nonoccurrence of an event as if they were assigned the numbers 0 and 1 and to run a standard ANOVA linear contrast. In particular, this method was discussed by W. Cochran in a survey paper he published a few years after Armitage's paper, in which Cochran was apparently ignorant of Armitage's prior publication. Thus the test is usually referred to as the Armitage–Cochran test.

Arvesen, J. N. (1969). Jackknifing U-statistics. *Ann. Math. Stat.* 40:2076–2100. This is a highly theoretical paper which establishes the most general conditions under which the pseudovariates of the jackknife tend to be independent and have a Gaussian distribution.

Barlow, R. E., Bartholomew, D. J., Bremner, J. M., and Brunk, H. D. 1972. *Statistical Inference Under Order Restrictions. The Theory and Application of Isotonic Regression*, Wiley, New York. This book gathers together all the various methods developed over the previous 10–15 years on this subject. By coincidence, the major workers in this field all had names beginning with B, and the book is sometimes referred to as B^4.

Burns, J. H. 1937. *Biological Standarizations*, Oxford University Press, London. Although a great deal of the material in this text is now outdated, Chapter II (which deals with estimating the LD_{50}) contains a great deal of good, solid common-sense comments on the use of mathematical models.

Cornell, J. A. 1981. *Experiments with Mixtures; Designs Models and the Analysis of Mixture Data*, Wiley, New York. This is a comprehensive work, bringing together all the activity in this field from the previous 20 years. Since mixture problems first arose in chemical engineering, much of the jargon is from engineering. However, the material is presented in a well-organized, concise fashion and can be used as an excellent introduction to this body of statistical techniques.

Cox, D. R. 1972. Regression models and life tables. *J. R. Stat. Soc. Ser. B 34*:187–220; 1975. Partial likelihood. *Biometrika 62*:269–276. The first paper established the methodology which is now used in the many computer programs that apply "Cox regression." There were some questions about the validity of the methodology, which Cox resolved in the more theoretical paper published in 1975.

Dempster, A. P., Selwyn, M. R., and Weeks, B. J. 1983. Combining historical and randomized controls in assessing trends in proportions. *J. Am. Stat. Assoc. 78*:221–227. This paper and that of Tarone (1982) both discuss Bayesian methods for combining historical data. Although both appeared within a year of each other, they passed like ships in the night. Each paper proposes a different (albeit Bayesian) approach and comes to a very different conclusion.

Dixon, W. J., and Massey, F. J. 1969. *Introduction to Statistical Analysis*, 3rd ed., McGraw-Hill, New York. The up-and-down method can be found in Chapter 19. This is an elementary "cookbook" type of text that is easy to read, where each chapter is self-contained, and with a large selection of statistical tables. It has proven to be an excellent reference book for the desk of many pharmacologists and toxicologists with whom I have worked. See Chapter 12 for analysis of covariance, Chapters 10 and 15 for a discussion of contrasts, and Chapter 13 for contingency table methods.

Dunnett, C. W., 1955. "A multiple comparison procedure for comparing several treatments with a control;" *J. Am. Stat. Assoc.* *50*:1096—1121.

ED-01 Task Force 1981. Re-examination of the ED 01 study. *Fund. Appl. Toxicol.* *1*:26—128. The material in these several papers was prepared by a task force of statisticians, toxicologists, and pathologists assembled by the Society of Toxicology to reexamine the data from a 24,000-mouse study run by the National Center for Toxicological Research in an attempt to delineate the pattern of low-dose response to a known carcinogen (2-AAF). Although some individual papers were not attributed to specific members of the task force, much of the material in the article entitled "Risk Assessment Using Time" (pp. 88—123) discusses proposals made by Robert Sielken, Jr., of Texas A&M dealing with the language of risk assessment. The paper entitled "Adjusting for Time on Study" discusses the failure of the data to fit the assumption of factorability of the hazard function.

Efron, B. 1969. Student's t-test under symmetry conditions. *J. Am. Stat. Assoc.* *64*:1278—1302. This is a highly theoretical paper which establishes sufficient conditions for the "robustness" of the t statistic to hold.

Elderton, W. P., and Johnson, N. L. 1969. *Systems of Frequency Curves*, Cambridge University Press, Cambridge. In the early years of the twentieth century Elderton published a complete description of the Pearsonian system of skew distributions. Johnson revised the text and brought it up-to-date with respect to other systems of continuous distributions.

Fisher, R. A. 1925. Theory of statistical estimation. *Proc. Cambridge Philos. Soc.* *22*:700—725. This is a full formal mathematical description of Fisher's theory of estimation. It is heavily theoretical but written with great clarity.

Fisher, R. A. 1925–1956. *Statistical Methods for Research Workers*, various editions, Oliver and Boyd, London. This is the classical "cookbook" text of Fisher in which he laid out the methods and ideas behind statistical analyses of controlled experiments. It is difficult to read, since methods of analysis have to be imputed from examples he gives and he makes no attempt to justify the techniques. However, it captures the spirit of Fisherian statistics, where the scientific conclusions are more important than the niceties of the mathematics.

Fix, E., Hodges, J. L., and Lehmann, E. L. 1959. The restricted chi-square test, in *Probability and Statistics (The Harald Cramer Volume)* (Grenander, ed.), Wiley, New York, pp. 92–107.

Food and Drug Administration, U.S. Department of Health and Human Services. *Guidelines for Toxicological Testing*, U.S. Government Printing Office, Washington, D.C. Various editions of these guidelines have appeared. They establish the acceptable protocols for toxicity tests for the FDA.

Hollander, M., and Wolfe, D. A. 1973. *Nonparametric Statistical Methods*, Wiley, New York. This book brings together most of the currently available methods in nonparametric statistics. It has just enough theory to give the flavor of the derivations but concentrates on the formulas and techniques used.

Kullback, S. 1968. *Information Theory and Statistics*, Dover, New York. Kullback's development of information theory represents a major theoretical structure for statistical inference. There are four or five of these fundamental structure, of which the Neyman–Pearson formulation of hypothesis testing is one. Most introductory tests in statistics approach statistical inference from the standpoint of decision theory (which is a generalization of the Neyman–Pearson work). Many of the resulting strictures on what is "proper" and what is not are based upon that theoretical structure. Users of statistics should be aware that there are other structures, with equally valid interpretations.

Litchfield, J. T., and Wilcoxon, F. 1949. A simplified method of evaluating dose-effect experiments. *J. Pharmacol. Exp. Ther. 96*:99–113. It is unfortunate that this paper is so often referenced but so seldom referred to. The graphical techniques proposed are easily done and they provide the toxicologist with a much better feel for how the data fit or do not fit the model than any abstract computer output.

Littlefield, N. A., Greenman, D. L., Farmer, J. H., and Sheldon, W. G. 1980. Effects of continuous and discontinued exposure to 2-AAF on urinary bladder hyperplasia and neoplasia. *J. Environ. Pathol. Toxicol.* 3:17—34. This paper is part of a group generated by the National Center for Toxicological Research from the 24,000-mouse ED-01 study (1981). Data from this study were also examined by a Society of Toxicology Task Force and additional papers published as indicated in ED-01 Task Force (1981).

Mantel, N. 1963. Chi square tests with one degree of freedom; extensions of the Mantel—Haenszel procedure. *J. Am. Stat. Assoc.* 58:690—700. This is a detailed discussion of a procedure published earlier by Mantel and Haenszel. Although the earlier paper is usually referenced by those using the procedure, this paper provides a complete discussion of the extensions that have since become an important part of the procedure.

Mantel, N., Bohidar, N. R., Brown, C. C., Ciminera, J. L., and Tukey, J. W. 1975. An improved Mantel—Bryan procedure for "safety" testing of carcinogenics. *Cancer Res.* 35:865—872. The initial Mantel—Bryan proposals appeared in two papers published in the late 1960s and in 1970. Problems quickly arose among those who tried to use it in situations where the control animals had tumors. This represents an attempt by Mantel and others to adjust the original proposals for more general cases.   .

Miller, R. G. 1974. The jackknife—A review. *Biometrika* 61:1—16. This is a thorough review of the theory and practices associated with jackknifing as of 1973 by one of the original workers in the field. Since then, literature on the jackknife has proliferated, owing mainly to its increasingly widespread use. However, this paper remains the most up-to-date and complete review available.

Morrison, D. F. 1967. *Multivariate Statistical Methods*, McGraw-Hill, New York. Multivariate statistical analysis is a forbidding subject to the uninitiated, requiring that the student be familiar with the very abstract theory of linear algebra. Of the many textbooks now available, this is probably the most accessible. Morrison introduces linear algebra as he needs it, sparingly, and gets directly to the statistical algorithms and the ideas behind them.

Neyman, J. 1935. Sur la vérification des hypothèses statistiques composées. *Bull. Soc. Math. Fr. 63*:246–266. This is a paper prepared by Neyman for readers unfamiliar with the current developments in statistics. It is written with great clarity and simplicity but clearly outlines the major ideas of and limitations of the Neyman–Pearson formulation.

Neyman, J., and Pearson, E. S. 1933. On the problem of the most efficient tests of statistical hypotheses. *Philos. Trans. A 231*:289–337. This paper is the culmination of a series of papers by Neyman and Pearson laying the foundations of their theory. It is highly mathematical but involves a very thorough discussion of the ideas and the limitations behind their formulation.

Salsburg, D. 1980. The effects of lifetime feeding studies on patterns of senile lesions in mice and rats. *Drug Chem. Toxicity 3*:1–33. This paper discusses the xylitol–sorbitol–sucrose study which provides data for some of the discussions in this text.

Savage, L. J., and Bahadur, R. R. 1956. On the non-existence of some statistical procedures. *Annals of Math. Stat., 27*: 1115-1122. In this paper, two of the most prominent theoreticians of the time combined to show that it is impossible to use statistical methods in a situation where the class of alternative hypotheses is too wide.

Tarone, R. E. 1982. The use of historical control information in testing for a trend in Poisson means. *Biometrics 38*:457–462. As indicated for Dempster et al. (1983), this is one of two papers published within a year of each other proposing different Bayesian solutions to the same problem.

Tarone, R. E., and Gart, J. H. 1980. On the robustness of combined tests for trends in proportions. *J. Am. Stat. Assoc. 75*:110–116. This paper is a full discussion, with examples from chronic toxicity studies, of the methods proposed by Tarone, which are, in turn, extensions of the Mantel–Haenszel procedure.

Wahrendorf, J. 1983. Simultaneous analysis of different tumor types in a long-term carcinogenicity study with scheduled sacrifices. *J. Nat. Cancer Inst.* 70:915–921. Data were taken from this paper to illustrate multidimensional table methods. Wahrendorf used log-linear models to analyze these data and reached a conclusion slightly different from the one proposed in this text.

Weil, C. S., and Wright, G. J. 1967. Intra- and interlaboratory comparative evaluation of single oral test. *Toxicol. Appl. Pharmacol.* *61m*:378–388.

Wermuth, N., and Lauritzen, S. L. 1983. Graphical and recursive models for contingency tables. *Biometrika* 70:537–552. This is a survey paper, albeit highly mathematical, that discusses the methods described in this text and which supplies an important set of references for anyone seeking to use these methods.

Index

Numbers in italics indicate detailed treatment of the subject.